Short Wave Listener's Guide

Short Wave Listener's Guide

Ian Poole

Newnes
An Imprint of Butterworth-Heinemann
Linacre House, Jordan Hill, Oxford OX2 8DP
A division of Reed Educational and Professional Publishing Ltd

A member of the Reed Elsevier plc Group

OXFORD BOSTON JOHANNESBURG
MELBOURNE NEW DELHI SINGAPORE

First Published 1997

British Library Cataloguing in Publication Data

A catalogue record for this book is available from the British Library.

ISBN 0 7506 2631 3

Library of Congress Cataloguing in Publication Data

A catalogue record for this book is available from the Library of Congress.

Typeset by Tekoa Graphics
Printed and bound in Great Britain by
Biddles Ltd, Guildford and King's Lynn

Contents

1 What is short wave listening?

Short wave listening is a hobby which has captivated the interest of many thousands of people over the years and today it is more popular than ever. There are many interesting stations which can be picked up from all over the world, and today's high technology receivers make it easier and more enjoyable to listen to them. From broadcast listening to amateur radio, experimentation to equipment construction, short wave listening is a hobby which can appeal to almost anyone.

1.1 Aspects of the hobby

Today a vast number of fascinating signals can be heard on the short wave bands. It is only necessary to tune a set across some of the bands to hear signals from all corners of the earth. Indeed part of the fascination of the hobby is that you never know where the next signal may come from. It may be a broadcast station from relatively close by. It may be a station from across the world, or possibly a radio amateur on a remote island.

One of the traditional areas of short wave listening is broadcast reception. There are thousands of stations located all over the world transmitting on the short wave bands. Programmes are in many languages, but English features very strongly, being spoken by a large proportion of the world's population. It is interesting to listen out for different broadcast stations. Sometimes their views are totally different from those heard on the normal domestic networks. In particular, radio stations are used extensively when there is a conflict. During the Gulf War many new stations were set up to broadcast propaganda or send out programmes for the troops. Listening to these stations can be a fascinating hobby in its own right.

Figure 1.1 *A short wave listening station (Courtesy of Peter Rayer G13038)*

Apart from broadcast stations, there are thousands of radio amateurs around the world. Even today this band of enthusiasts is contributing to our knowledge about different aspects of radio including the way radio signals travel around the earth and new methods of sending data. That apart, many amateurs enjoy the hobby as a form of relaxation and can be heard chatting to one another about all manner of interesting subjects.

Sometimes radio amateurs have provided an invaluable service to others by establishing the only means of communication from a disaster area. Islands cut off by hurricanes have depended upon radio enthusiasts to maintain links with the outside world. Similarly, in war torn countries amateur enthusiasts have saved many lives by organizing urgent medical supplies or informing the outside world of what has been happening. Sometimes transmissions of this kind can be picked up, adding a new dimension to short wave listening.

There are many other interesting signals to be heard as you tune across the short wave bands. Maritime and aeronautical services use the short wave bands. Even though satellite communications are becoming more common, short wave communications are still indispensable. You may pick up signals from all parts of the earth forming part of these long distance communications services.

1.2 Technology today

The performance of the receiver is of paramount importance for any short wave listener. Today receivers are benefiting from the latest advances in technology. Their radio frequency circuits are using new devices to achieve the required levels of sensitivity, whilst other developments are enabling the selectivity, the ability to reject unwanted signals, to be improved, reducing the level of interference.

Recent developments in computer technology have made possible new facilities for short wave receivers. For example, some now offer keypad entry of frequencies and memories to store the frequencies of favourite stations. Some sets can be remotely controlled from a computer.

Figure 1.2 *A modern communications receiver (Courtesy Yaesu UK Ltd)*

A wide variety of receivers is available nowadays. Some portable radios are designed with short wave reception in mind. Using today's technology they offer a good performance on the short wave bands as well as the long, medium and VHF FM bands provided by normal portables. For those specifically interested in short wave coverage, more sophisticated communications receivers offer near professional performance at competitive prices.

It is not only receiver design that has been affected by computer technology. New modes of transmission have been introduced. While the traditional modes, such as Morse and voice or sound transmission, have remained unchanged for decades, many developments have occurred recently in the area of data communications. For example, the old teleprinter transmission systems, so unreliable and prone to errors caused by interference, are being replaced by new forms of data transmission developed in the computer industry. These are more versatile and they include error-correcting procedures. If errors are detected, the data is re-sent automatically until it is received correctly.

1.3 Looking to the future

At one time people thought that the short waves would be totally superseded by satellite communications. This view is not held by many now. The short wave bands remain a very useful medium for long distance transmissions and are much cheaper than satellite communications.

New developments are taking place in a number of areas. In broadcasting, pressure of space is as fierce as ever, as evidenced by the allocation of additional space to broadcasters recently. Moreover, it has been agreed that broadcasters will convert to single sideband, a form of transmission which occupies less space. Naturally this cannot be accomplished overnight because many radio receivers will not be able to cope with these changes.

Whilst it is still possible to listen to transmissions using the older modes of communication, the host of new developments show that the short wave scene is as fast-moving as any area of radio technology.

1.4 How it began

While it is right that a technological hobby like short wave listening should look to the future, it is also interesting to take a quick look at how the hobby began and has grown over the years. From this it is possible to see some of the reasons for present practices.

The foundations for the hobby were laid when a brilliant Scot named James Maxwell mathematically proved the existence of an electromagnetic wave which would be able to travel over immense distances. Unfortunately Maxwell was never able to perform any practical experiments to confirm his theory.

The honour of physically proving the existence of electromagnetic waves fell to a German named Heinrich Hertz. He undertook a number of experiments which proved the existence of these waves beyond all doubt. In one he used an induction coil connected to a loop of wire. Two spheres were placed in the loop with a small gap between them. When the induction coil generated a large voltage, a spark jumped across the gap between the two spheres. Simultaneously a smaller spark was seen to jump across a gap in a similar loop placed a small distance away.

It was not until some time after the discovery of these waves that people realized their benefits for communications. A young Italian named Guglielmo Marconi was one of the first to see these possibilities.

Marconi began by repeating Hertz's experiments. Gradually he increased the distances over which he could detect signals. Although he had a natural flair for the subject, in those pioneering days his improvements relied mainly on trial and error.

When he had a system which was capable of operating over a useful distance, he decided to investigate the commercial possibilities. First Marconi approached the Italian Ministry of Posts, but he was rejected. So he emigrated to England where his mother had contacts with the scientific community.

In England Marconi developed his system further, each improvement, such as his sending messages across the English Channel, attracting more publicity. However, it was his achievement in 1901 of sending a signal across the Atlantic that made Marconi an instant hero and won worldwide publicity for this new method of communication.

Amateur experimenters took a keen interest in the subject. Home model makers' and constructors' magazines of the time showed people how to make their own equipment. In those early days of wireless there was no need for licences for transmitting or receiving, but as interest grew the government saw that some form of regulation was needed. Accordingly licences for amateur experimentation were first issued in the UK in 1904.

The number of licences grew steadily although they were revoked during the First World War. After the war, interest in the hobby grew rapidly and amateurs began to make radio history. Greater distances were covered and soon the first short wave transatlantic contacts were made. These differed from Marconi's contacts in that he used the long wave band. Amateurs were not allowed access to these wavelengths as they were used for commercial traffic. Instead they were confined to the short wave bands which were thought to be of little practical use.

The start of broadcasting in 1919 stimulated rapid growth of interest in radio in the early 1920s. Many broadcasting stations came on the air and the British Broadcasting Company was formed in 1922, taking over transmissions from a number of independent operators. In 1927 it became the British Broadcasting Corporation, the BBC we know today.

The rising number of broadcasts brought about a corresponding increase in the number of sets being bought. Because wireless sets were very expensive then, many people made their own. Consequently, a burgeoning industry grew up supplying the parts needed by the growing number of radio constructors.

With national broadcasting established, people started to turn their eyes to international broadcasting on the short waves, early transmissions being made in 1925. The BBC started experimental short wave transmissions in 1927 and introduced a permanent service in 1932. The aim of their 'Empire' service was that every country in the British Empire should receive a broadcast from London at least once each day.

Other countries also started short wave broadcasts. Their power as an instrument of foreign policy was soon recognized. Italy and Germany both used international broadcasting in the Second World War. Many broadcasters used it in the Cold War and more recently it was again widely used in the Gulf War of 1991.

All these developments served to establish listening to the short waves as a fascinating hobby. Broadcast listening and listening to radio amateurs are accepted as interesting pastimes. Today there is an even wider variety of transmissions that can be received, although listeners should be aware of laws governing which transmissions can be legally monitored.

2 Radio transmissions

Most radio signals are used to carry information. They are used for a variety of purposes, from carrying audio for sound broadcasting or point-to-point communications to carrying data in a variety of forms including telexes and pictures.

To transmit any form of information by radio, a signal or **carrier** is first generated. The information, whether audio or some other form of data, is used to modify or **modulate** the carrier. In this way the information is superimposed onto the carrier which is transmitted to the receiver. Here the information is extracted from the radio signal and reconstituted in its original form, a process known as **demodulation**.

There are many ways in which a carrier can be modulated. Each has its own advantages and can perform well under given conditions. Some of the simpler forms of transmission have the advantage that they can be demodulated by quite simple receivers. On the other hand, those modes which need more complicated circuitry to resolve them may enable reception under poor conditions, such as when interference levels are high or signal levels are low.

2.1 Radio carrier

The basis of any radio signal or transmission is the carrier. This consists of an alternating waveform like that shown in Figure 2.1. The rate at which the signal alternates is known as the **frequency**, and it is the number of cycles which are completed within a given time. Normally this time is one second, and a frequency of one cycle per second is known as 1 hertz, named after the German scientist Heinrich Hertz. The frequencies which are usually found in radios are very much greater than this. Prefixes are added to indicate multipliers. 1 kilohertz (kHz) is

1000 hertz, 1 megahertz (MHz) is 1 000 000 hertz, and 1 gigahertz (GHz) is 1 000 000 000 hertz.

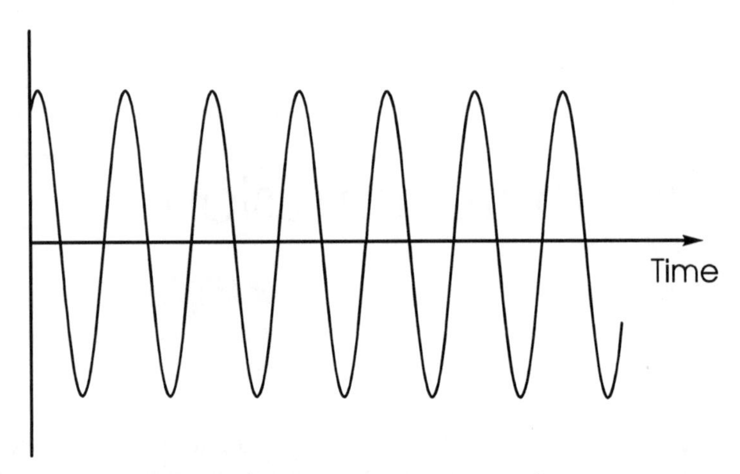

Figure 2.1 *An alternating waveform*

2.2 Modulation systems

The carrier in itself does not convey any information. It is the pure radio signal; the information it is to carry is used to modulate it. There are many modulation systems; some are complicated, requiring complex circuitry to encode and decode them. Others are very much simpler.

2.2.1 Morse

Morse is the oldest and simplest way of transmitting information using radio. Despite its age it still has several advantages over other forms of transmission, and this means that it is still in fairly widespread use today.

One of its advantages is its simplicity. It only consists of a carrier wave which is turned on and off as shown in Figure 2.2. The characteristic dots and dashes are defined by the length of time the transmission is left on. The dots and dashes then make up the required letters according to the Morse Code given in Table 2.1. From this it can be seen that the signal transmitted in Figure 2.1 is a dash and a dot which makes the letter 'A'.

The code has been used in this form for over a century. The first code devised by Samuel Morse had a number of shortcomings, so he revised it and it has been used in this form ever since. As it is used worldwide, it is often called the International Morse Code. This distinguishes it from other codes used by countries like Russia which have a different alphabet.

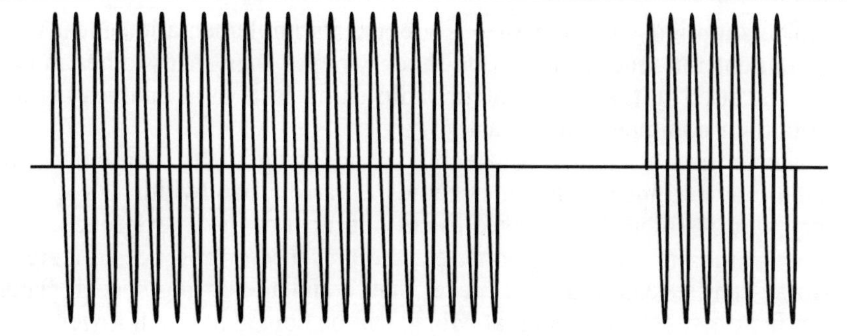

Figure 2.2 *A Morse signal*

Table 2.1 The Morse Code

A	·—	N	—·	1	·————
B	—···	O	———	2	··———
C	—·—·	P	·——·	3	···——
D	—··	Q	——·—	4	····—
E	·	R	·—·	5	·····
F	··—·	S	···	6	—····
G	——·	T	—	7	——···
H	····	U	··—	8	———··
I	··	V	···—	9	————·
J	·———	W	·——	0	—————
K	—·—	X	—··—		
L	·—··	Y	—·——		
M	——	Z	——··		

Punctuation

Full Stop	·—·—·—	Equals sign (=)	—···—
Comma	——··——	Stroke (/)	—··—·
Question Mark (?)	··——··	Mistake	········

Procedural Characters

Procedural characters made up of two letters are sent as a single letter with no break between them

Start of Work (CT)	—·—·—	End of Message (AR)	·—·—·
Invitation to Transmit (KN)	—·——·	Invitation to Transmit (K)	—·—
End of Work (VA)	···—·—	Invitation to a Particular Station to Transmit (KN)	—·——·

Timing

Length of a dash	3 dots	Space between letters	3 dots
Space beteen dots and dashes	1 dot	Space between words	5 dots

The simplicity of the Morse Code and its implementation means that equipment for sending it can be much simpler than many other modes. This is an advantage for amateur radio enthusiasts as it enables them to build their equipment more easily.

Morse has a number of technical advantages. Its relatively slow signalling rate means that it occupies a narrow bandwidth. As a result very narrow filters can be used to cut out most of the interference. Also, because the brain has only to detect the presence or absence of a signal, Morse can be read at a lower level than a signal carrying speech. These two factors together mean that Morse can be copied at much lower signal levels than other forms of transmission.

To send a message a Morse key is needed to turn the transmission on and off. A typical key is shown in Figure 2.3. It is essentially a switch. When pressed down contact is made and the transmission is turned on. Releasing the pressure allows the spring-loaded arm to return, breaking the contact and turning the transmission off. Using a key like this it is possible to send messages at a speed of up to 20 words per minute or thereabouts. Some people can send much faster than this, but it can become tedious over long periods of time.

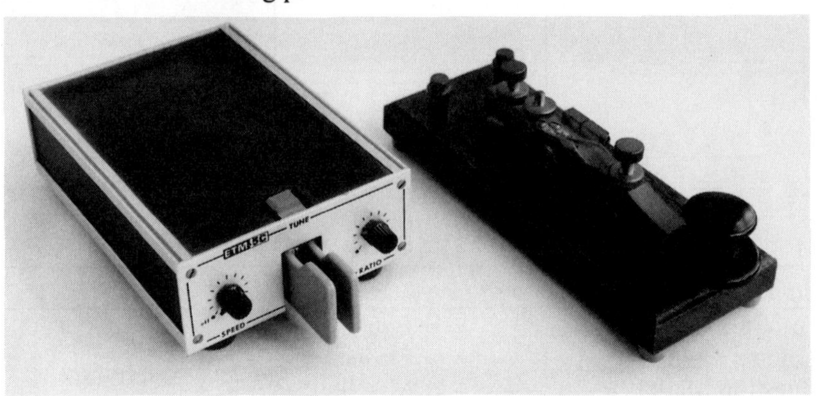

Figure 2.3 *A Morse key and keyer*

To send Morse at high speed, electronic keyers like the one shown in Figure 2.3 are often used. The two paddles at the front of the unit are used for sending. If the left hand one is pressed inwards a series of dots is generated. If the right hand one is pressed inwards, dashes are generated. If both are squeezed together then dots and dashes are produced alternately. Although a little practice is needed to master a keyer like this, it does make sending good Morse at high speed much easier.

Computers can also be used for generating Morse. A number of programs are available: the operator simply types the message on the keyboard and the computer generates the corresponding Morse Code. Computer techniques can also be used for decoding Morse signals. The audio output from the receiver can be fed into a special demodulator and the computer deciphers the message. Whilst automatic decoders work well when good Morse is sent and conditions are ideal, they soon start to fail when interference levels rise and signal levels fall. The human ear and brain are far superior to even the latest automatic decoding systems.

The most common way to send Morse is simply to interrupt the carrier wave. However, if this is picked up by an ordinary domestic portable radio receiver it will simply be heard as a series of clicks and pops as the signal is turned on and off. To make the characteristic tone, a circuit called a **beat frequency oscillator** (BFO) (explained in Chapter 4) is used. An alternative practice is to modulate the carrier with an audio tone which is turned on and off. Although this is a perfectly viable method of sending Morse, it is not as efficient as simply turning the carrier on and off. For this reason the method is seldom used.

2.2.2 Amplitude modulation

Whilst Morse Code has its advantages, we have all become accustomed to hearing music and speech over the radio. There are a number of ways in which a carrier can be modulated to take an audio signal. The most obvious way is to change its amplitude in unison with the variations in intensity of the sound wave. In this way the overall amplitude or envelope of the carrier is modulated to carry the audio signal as shown in Figure 2.4. Here the overall amplitude or envelope of the carrier can be seen to change in line with the modulating signal.

Amplitude modulation or AM is one of the most straightforward methods of modulating a signal. Demodulation, the process whereby the audio signal is recovered from the radio-frequency signal, is also simple. It requires only a diode detector circuit like that shown in Figure 2.5. In this circuit the diode rectifies the signal, only allowing the one half of the alternating radio frequency waveform through. A capacitor removes the radio-frequency component of the rectified signal, leaving a pure audio signal. This can be fed into an amplifier to drive a loudspeaker. As the circuitry for demodulating AM is so simple, it helps to keep the cost of AM receivers low.

Whilst AM has its advantages of simplicity, it is not the most efficient transmission mode to use, either in terms of the amount of spectrum it takes up or in usage of transmitter power. For these reasons it is rarely used for communications purposes. Its only major communications use is for VHF aircraft communications. It is still widely used, however, on the long, medium and short wave bands for broadcasting, because of its simplicity.

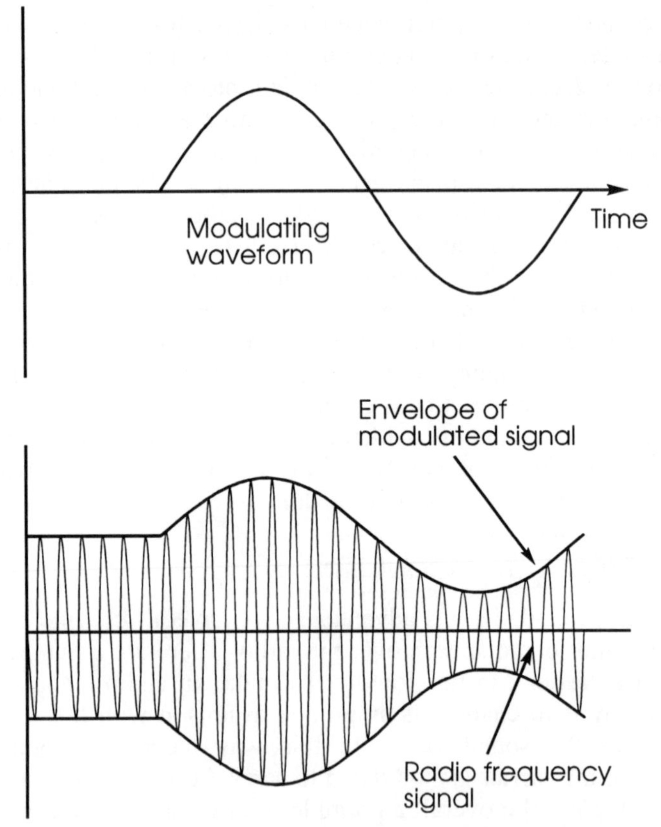

Figure 2.4 *Amplitude modulated signal*

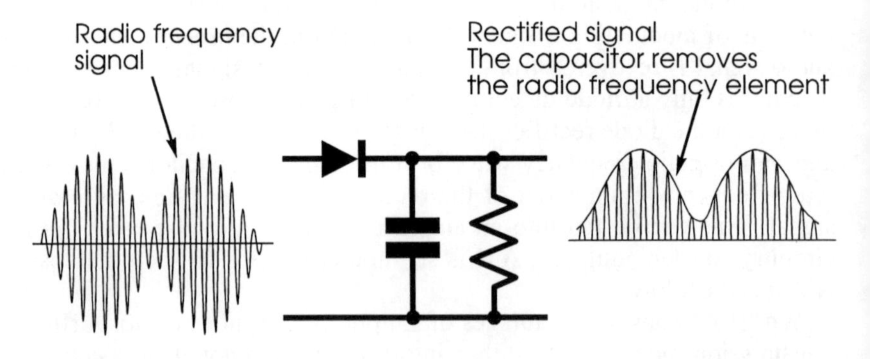

Figure 2.5 *A simple diode detector*

To find out why it is inefficient it is necessary to look at the theory behind the operation of AM. When a radio frequency signal is modulated by an audio signal the envelope will vary. The level of modulation can be increased to a level where the envelope falls to zero and then rises to twice the unmodulated level. Any increase on this will cause distortion. As this is the maximum amount of modulation possible it is called **100% modulation**.

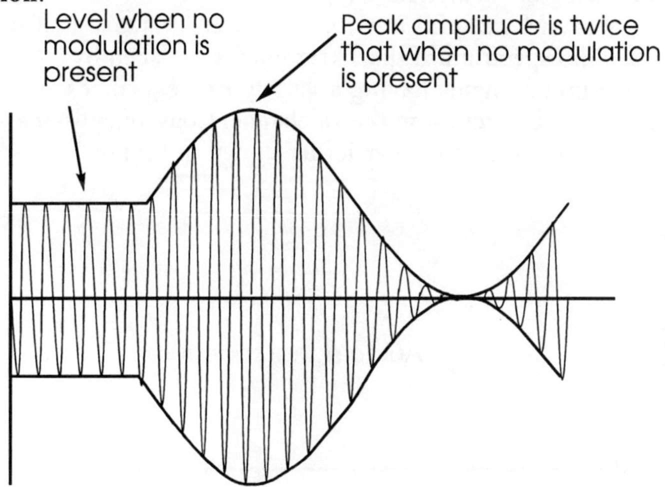

Figure 2.6 *Fully modulated signal*

Even with 100% modulation the utilization of power is very poor. When the carrier is modulated, **sidebands** appear at either side of the carrier, and it is these that contain the information about the audio modulation. To see how the signal is made up and the relative power distribution within it, consider the simple example of a 1 kHz tone modulating the carrier. Two signals will be found 1 kHz either side of the main carrier, as shown in Figure 2.7. When the carrier is being fully

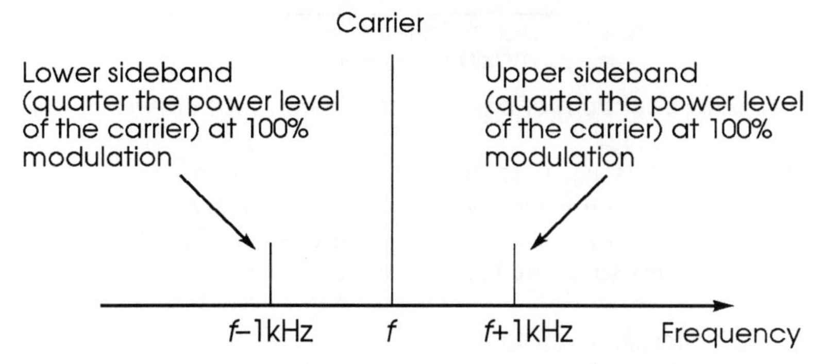

Figure 2.7 *Spectrum of a signal modulated with a 1 kHz tone*

modulated, i.e. 100%, the power in each sideband rises to half the voltage level of the main carrier. In terms of power this means that each sideband is just a quarter of the main carrier's level. During the modulation process the carrier remains constant and is only needed as a reference during the demodulation process.

Not only is AM wasteful in terms of power, it is also inefficient in its use of the radio spectrum. If the 1 kHz tone is replaced by a typical audio signal made up of sounds having a variety of frequencies, each of those frequencies will be present in the sidebands. Consequently the sidebands spread out either side of the carrier as shown in Figure 2.8 and the total

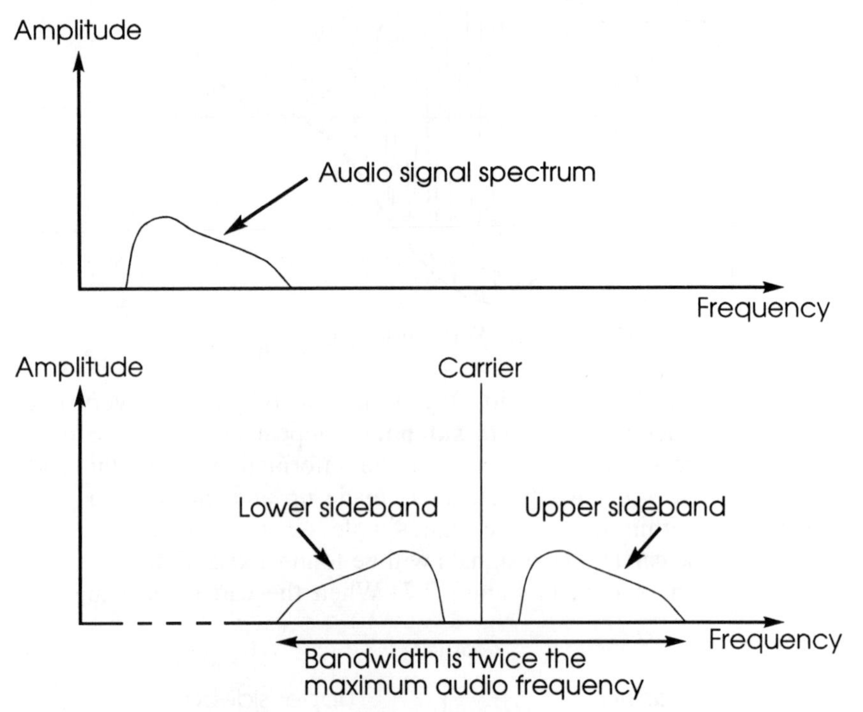

Figure 2.8 *Spectrum of a signal modulated with speech or music*

bandwidth used is equal to twice the top audio frequency which is broadcasted. In the crowded conditions found on many of the short wave bands today, this is a waste of space, and other modes of transmission which take up less space are being used increasingly.

2.2.3 Single sideband

One of the modes widely used for communications traffic is called **single sideband** (SSB). This is a derivative of AM in which the signal is manipulated to overcome the disadvantages of AM and give a more efficient mode of transmission.

There are two main stages in the generation of a single sideband signal. The first is that the carrier is removed. Because it does not contribute to carrying the sound information and is only used during demodulation, it is possible to remove it in the transmitter, as shown in Figure 2.9, enabling power to be saved.

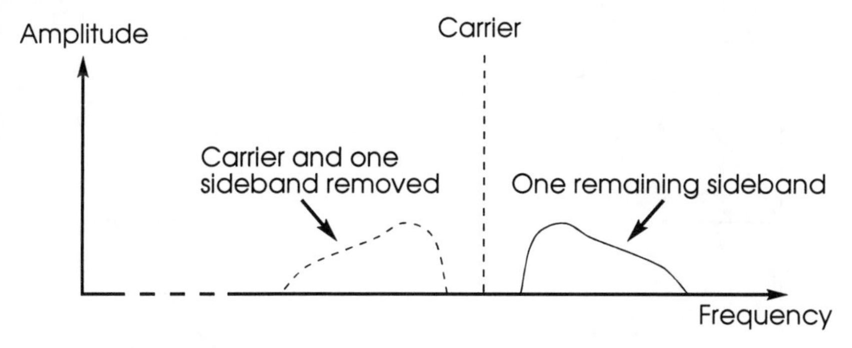

Figure 2.9 *Spectrum of a single sideband signal*

Moreover, only one sideband is needed. The upper and the lower sidebands are mirror images, so either could be used for conveying the audio information. Having only one sideband halves the bandwidth used, without degrading the signal. Another advantage of transmitting only one sideband is that filter bandwidths in the receiver can be made narrower to reduce interference and give better reception.

To demodulate the signal the carrier has to be reintroduced in the receiver using a **beat frequency oscillator** (BFO), also called a **carrier insertion oscillator** (CIO). The BFO must be on the correct frequency relative to the sideband being received. Any deviation from this would affect the pitch of the recovered audio. Whilst errors of up to about 100 Hz are acceptable for communications purposes such as amateur radio, the reintroduced carrier must be maintained on exactly the correct frequency if music is to be transmitted. This is accomplished by transmitting a carrier at a very low level. Circuitry in the receiver locks on to this.

As either sideband could be used, a convention determines which is used. In this way the receiver can be set up to expect the received sideband. If this is not done then the receiver has to be continually switched between upper and lower sideband. Commercial operators use the upper sideband in all instances, whereas radio amateurs use the upper sideband on bands above 10 MHz and lower sideband below this.

2.2.4 Frequency modulation

The most obvious method of applying modulation to a signal is to superimpose the audio signal onto the amplitude of the carrier. This, however, is by no means the only method which can be employed. It is also possible to vary the frequency of the signal to give **frequency modulation** or FM. Figure 2.10 shows how the frequency of the signal varies with the modulating voltage.

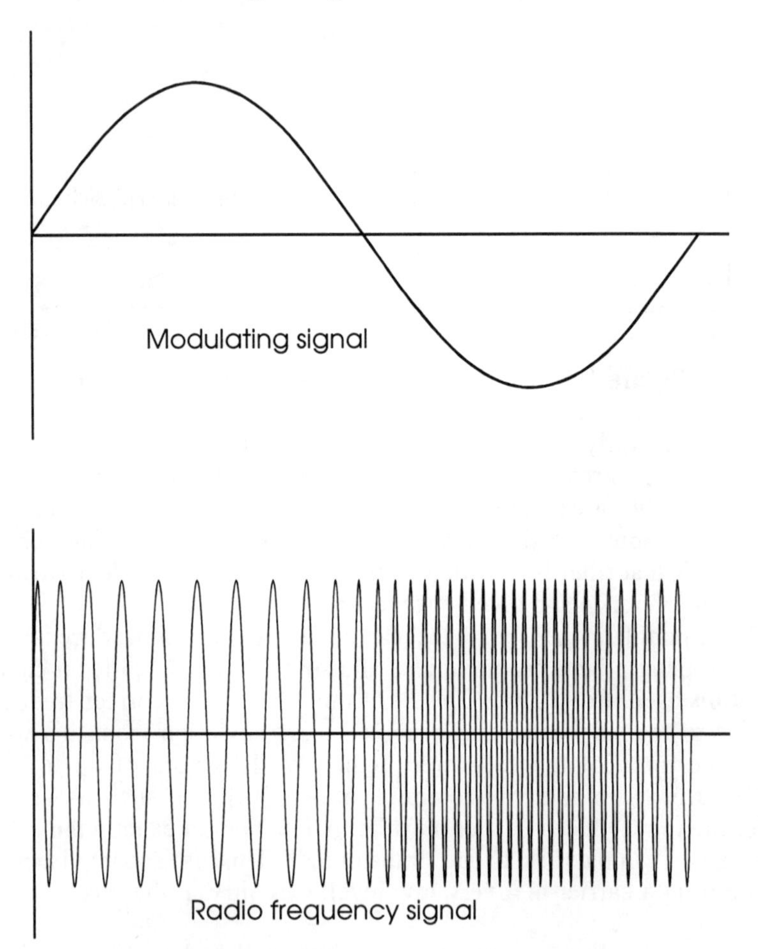

Figure 2.10 *A frequency modulated signal*

The amount by which the signal frequency varies is very important. This is known as the **deviation** and is normally quoted as the number of kilohertz deviation. For example, if the signal has a deviation of ±3 kHz, the carrier frequency varies by 3 kHz.

Broadcast stations in the VHF portion of the frequency spectrum between 88.5 and 108 MHz use large values of deviation, typically ±75 kHz. This is known as **wide-band FM** (WBFM). These signals are capable of supporting high-quality transmissions, but they occupy a large bandwidth. Usually 200 kHz is allowed for each wideband FM transmission. For communications purposes less bandwidth is used. This **narrow-band FM** (NBFM) typically uses deviation figures of ±3 kHz.

FM is used for a number of reasons. One advantage is its resilience to signal level variations. Because the modulation is carried only as variations in frequency, signal level variations will not affect the audio output, provided that the signal does not fall to a level where the receiver cannot cope. Consequently FM is ideal for mobile or portable applications where signal levels are likely to vary considerably.

Another advantage of FM is its immunity to noise. Most noise appears in the form of amplitude variations. This will be ignored in the receiver if it detects only frequency variations. This is why FM is used for high-quality broadcast transmissions where low background noise levels are of paramount importance.

To demodulate an FM signal it is necessary to convert the frequency variations into voltage variations. This is slightly more complicated than demodulating AM, but it is still relatively simple. Instead of rectifying the carrier using a diode, the demodulator is designed to produce an output voltage proportional only to the carrier frequency deviation and to ignore variations in carrier amplitude.

2.2.5 Frequency shift keying

It is not easy to modulate digital data directly on to a radio carrier. The square edges of the data require a wide bandwidth, and the transmission path is likely to distort the signal so that large numbers of data errors are encountered. To overcome this a different approach is adopted.

Many signals heard on the bands employ a system called **frequency shift keying** or FSK. The signal alternates between two frequencies, one representing the digit one (sometimes called the **mark**) and the other the digit zero (the **space**). By switching between these two frequencies it is possible to send digital data over the radio.

FSK is widely used on the HF bands. To generate the audio tone required from the receiver, a beat frequency oscillator must be used. For this reason to obtain the correct audio tones the receiver must be tuned to the desired frequency with very high accuracy.

At frequencies in the VHF and UHF portion of the spectrum a slightly different approach is adopted. An audio tone is used to modulate the carrier and the audio is shifted between the two frequencies. Although the carrier can be amplitude modulated, frequency modulation is virtually standard. Audio frequency shift keying (AFSK) makes the tuning of the receiver less critical.

When the data signal leaves the receiver it is generally in the form of an audio signal switching between two tones. This is converted into the two digital signal levels by a unit called a **modem** which stands for MOdulator/DEModulator. Audio tones fed into the receiver will generate the digital levels required for a computer or other equipment to convert into legible text. Conversely it can convert the digital signals into the audio tones required to modulate a transmitter sending data.

The speed of the transmission is important. For the receiver to decode the signal correctly it must know the rate at which it is arriving. To facilitate this a number of standard speeds are used. These are normally given as a certain number of **baud**. One baud is equal to one bit per second.

2.2.6 Radio teletype

The first data transmissions used large noisy mechanical teleprinters or teletypes for a mode known as **radio teletype** or RTTY for short. The machines converted the switched signal voltages from the receiver into printed output on paper. These machines were widely available on the surplus market for many years and they were pressed into service for listening as well as transmitting. They are not often used now because computers can perform the same function far more conveniently, but RTTY transmissions are still widely used.

RTTY uses a code of five pulses sent one after another to represent the letters, figures and other symbols which need to be transmitted. The code which is used is called the Murray or Baudot code and is shown in Table 2.2. This code is internationally recognized and to accommodate the various international requirements some national variations exist.

Data is sent relatively slowly because the mechanical teleprinters could not cope with data any faster. 45.5 and 50 baud were the two main standards used, although other standards at 56.88, 74.2 and 75 baud exist. For amateur use the standard speed used for transmissions is 45.5 baud.

A frequency shift of 170 Hz is widely used. A frequency of 1445 Hz represents the mark condition and 1275 Hz gives a space.

The letter shift and figure shift codes are sent to change from upper case to lower case and vice versa. Once the code has been sent then the system will remain in that case until the next case change code is sent.

One of the major problems with RTTY is that any interference causes the received data to be corrupted. Even under relatively good conditions it is very difficult to have a totally correct copy. Today's new data modes utilize the power of computers to detect and correct errors.

Table 2.2 Murray or Baudot Code

Lower Case	*Upper Case*	*Code Element*					*Decimal Value*
		5(MSB)	*4*	*3*	*2*	*1*	
A	–	0	0	0	1	1	3
B	?	1	1	0	0	1	25
C	'	0	1	1	1	0	14
D	$ AB	0	1	0	0	1	9
E	3	0	0	0	0	1	1
F	! %	0	1	1	0	1	13
G	& @	1	1	0	1	0	26
H	□	1	0	1	0	0	20
I	8	0	0	1	1	0	6
J	' Bell	0	1	0	1	1	11
K	(	0	1	1	1	1	15
L	)	1	0	0	1	0	18
M	.	1	1	1	0	0	28
N	,	0	1	1	0	0	12
O	9	1	1	0	0	0	24
P	0	1	0	1	1	0	22
Q	1	1	0	1	1	1	23
R	4	0	1	0	1	0	10
S	Bell !	0	0	1	0	1	5
T	5	1	0	0	0	0	16
U	7	0	0	1	1	1	7
V	; =	1	1	1	1	0	30
W	2	1	0	0	1	1	19
X	1	1	1	1	0	1	29
Z	" +	1	0	0	0	1	17
Space		0	0	1	0	0	4
CR		0	1	0	0	0	8
LF		0	0	0	1	0	2
Figure Shift		1	1	0	1	1	27
Letter Shift		1	1	1	1	1	31
Blank		0	0	0	0	0	0

AB = Answer back or WHU (Who Are You?)

Upper case characters may vary in some cases as indicated on the chart. Also upper case F, G and H are not often used.

2.2.7 SITOR and AMTOR

To overcome the problems with RTTY a commercial system known as **SITOR** was developed. This is widely used by the maritime services where it gives more reliable communication. It gives improved performance by using a coding system which allows errors to be detected and corrected. The amateur radio version called **AMTOR** is very similar and is widely used for communications on the HF bands. These systems send data at 100 baud.

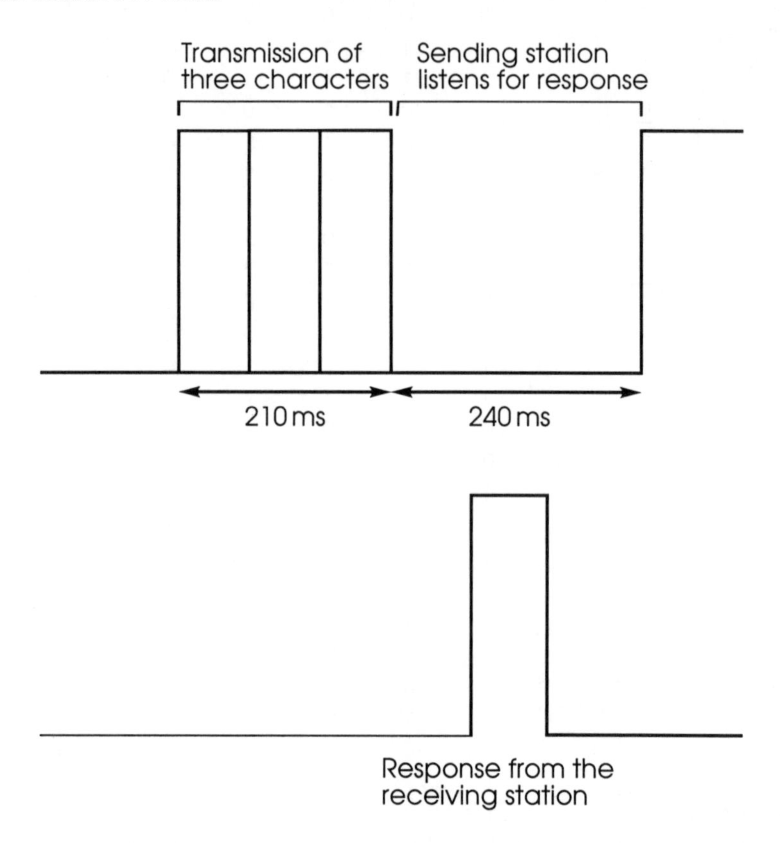

Figure 2.11 *Timing of an Amtor transmission in mode A*

The system uses the same basic 5-bit code for sending the data. However, a total of seven bits is sent and the additional bits are used in such a way that the transmitted data pattern always contains four mark bits and three space bits. From a knowledge of this expected pattern the receiver is able to detect an error and take action to correct it.

In operation the transmitter sends out three characters. The receiver checks them to ensure they are correct. If they are, then a reply is sent back to indicate they have been received correctly. Then the next block of three characters can be sent out. If they have not been received correctly then this is indicated and the block is re-sent.

A block takes a total of 450 ms to send. Each character takes 70 ms giving a total of 210 ms for the transmission. Then there is a window of 240 ms for an acknowledgement to be received. This amount of time is allowed to take account of the delays which occur.

The method of sending which uses an automatic request for repeat (ARQ) is called **mode A**. It can only operate if contact has been established with a particular station. A general transmission such as a news bulletin or an amateur radio operator wanting a contact cannot work in these situations. For these a second mode, called **mode B,** is used. In this mode each character is sent twice. Initially the first character is sent once and then the repeat message is sent five characters behind the first one. The time interval between the two signals reduces the possibility of interference causing problems. Sending the data twice also gives the receiver two attempts at capturing each character. Moreover, because seven bits of data are sent instead of the five used for the character code itself, error detection is still possible, allowing the receiving equipment to decide which character of the two to accept.

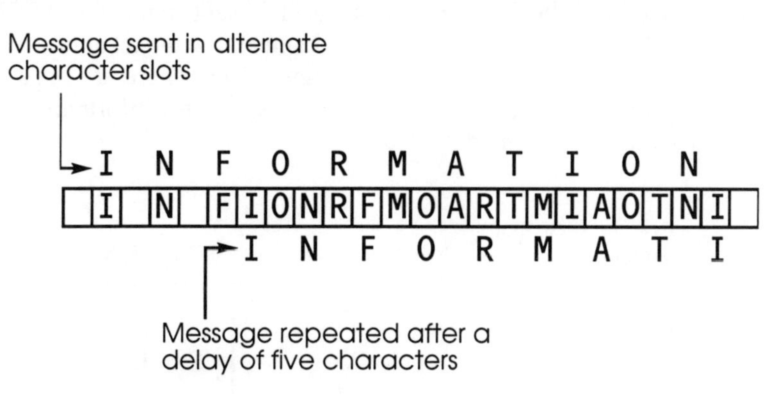

Figure 2.12 *Amtor mode B transmission*

2.2.8 *Packet*

Packet radio is widely used on the amateur bands, particularly on frequencies above 30 MHz where it is one of the fastest growing forms of communication. The basic system uses computer technology to enable error-free communication combined with many useful facilities.

As the name implies, this mode of transmission splits the data into a series of **packets** sent one at a time. Messages are usually too long to be sent in one packet, so they may take take several packets.

One of the advantages of packet radio is that one channel can be used by many stations. This means that when sending data a station has to wait until the channel is clear. When the frequency is free, the first packet can be sent, and the receiving station will return an acknowledgement that the data has been received correctly. If this is not received, the transmitting station waits for the frequency to clear and re-sends the data. This process is repeated until the data has been correctly received. Once the first packet has been transferred, the second and subsequent ones are all transmitted in the same way. As the receiving station checks for errors and the transmitter repeats the data until it has been correctly received, the system is very resilient and gives high accuracy. The method of waiting until the frequency is clear before transmitting allows many stations to use the same frequency, providing efficient utilization of the available spectrum.

Like other data modes packet radio uses frequency shift keying. A transmission speed of 1200 baud with tone frequencies of 2200 Hz for the space and 1200 Hz for the mark condition has been adopted for VHF. On HF where conditions are a little more difficult, a speed of 300 baud with a 200 Hz shift is generally employed.

The format for each data packet is accurately defined so that the receiver can decipher the incoming data. Data is sent using the ASCII code shown in Table 2.3 (at the end of this chapter). Each packet has five different elements or sections as shown in Figure 2.13. There are flags at the beginning and end of each packet, an address, control information, a frame check sequence and the data itself.

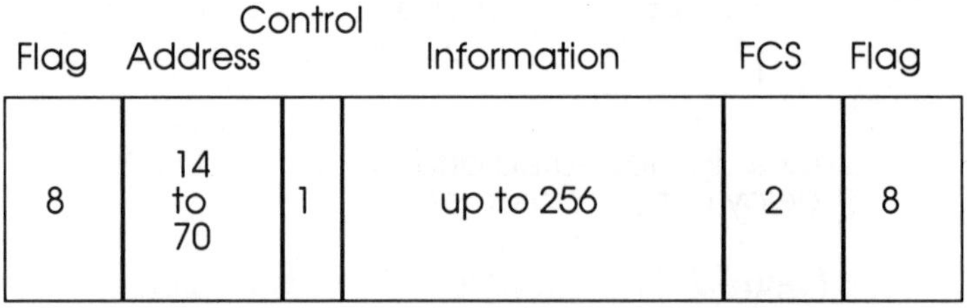

Flag	Address	Control	Information	FCS	Flag
8	14 to 70	1	up to 256	2	8

Figure 2.13 *Format of a packet transmission*

The flag at the beginning of the packet is used to allow the receiving decoder to synchronize to the incoming data. This is followed by a station address. This defines the callsign of the station to which the data is being sent. Also included is the source or sending station callsign and the callsigns of any repeaters which are to be used to relay the message. This means that any other station using the frequency will be able to ignore the data and only receive the signals intended for it.

The next piece of information to be sent is the control byte. This is used to signal acknowledgements and requests to repeat transmissions. This is followed by the data itself. The length of this can be up to 256 bytes. Once it is complete the frame check sequence or FCS is sent. This is a check-sum whose value is calculated from the data. It is used by the receiving station to check that all the data has been correctly received. Only when the receiver is able to generate a code to match the received one is an acknowledgement sent.

The final part of the packet is the terminating flag. This is recognized by the receiver as the end of the message and enables it to check the data and send its acknowledgement.

Packet radio is able to utilize a number of features which were not present in previous types of data communication. One of the most widely used is the ability for other stations to relay messages, so that much greater distances can be covered. Stations which relay messages in this way are called digital repeaters or **digipeaters** for short.

Packet radio transmissions take place on a single frequency. This means that digipeaters have to receive and transmit on a single frequency. For them to be able to relay the messages, the message must first be received in full, stored and then transmitted. Once the final station in the chain has received the message, the acknowledgement is sent back along the chain to the first station. This is known as an **end-to-end acknowledgement**. Only then is the next packet sent. This means that when a message is sent over a long path using several stations as repeaters, the message can take a long time to get through, especially if packets have to be repeated.

One powerful facility which packet radio offers is the ability to read data from a mailbox. Sometimes called a **bulletin board system** (BBS), it enables messages to be sent to a particular mailbox and left for collection by a particular station. In many respects it is like a radio e-mail system. A message is sent to the local mailbox. Once received it is stored and then it is passed on via a network of mailbox stations until the required destination mailbox is reached. The message is stored at this mailbox until it is read by the recipient station. The advantage of using the mailbox system is that it is not necessary to know the route required to be taken by the message. This is worked out by the system, as it has a knowledge of the stations and works out a suitable route. Data is generally sent at periods of low activity, often at night, and this means that messages can take a few days to arrive. However as many links exist between countries it is possible to send messages around the world. In addition to the basic mail facility, many items of general interest are stored and can be accessed by any station.

Although much of the initial experimental and practical work regarding packet radio was carried out by radio amateurs, many commercial packet systems are now in use around the world. They are particularly useful where a large number of users have to send small amounts of data at intervals which would not demand a separate frequency for each separate user. For example, packet radio is ideal for monitoring systems where each outpost has to be polled or accessed at intervals, or where it periodically reports a status or other information to the main station.

To be able to decode AMTOR, SITOR and packet, equipment has to be connected to the receiver. A modem is the basic requirement together with a controller and display unit or printer. It is not always necessary to use a controller as a computer can perform this function as well as displaying any messages.

2.2.9 PacTor

This mode takes the best features from packet and AMTOR to improve the transmission of data, particularly on the HF bands. The relatively large amounts of data sent in each packet transmission can be difficult to receive correctly in poor conditions. On the other hand AMTOR is relatively slow even when conditions are good, and much higher data rates could be achieved with little difficulty. PacTor aims to overcome these problems by adapting to the prevailing circumstances.

PacTor operates at two different speeds according to the conditions. Under normal conditions the system operates at 100 baud, but it swaps to 200 baud if a good link is established. In the slow speed each packet of data is 14 bytes long. This takes a total of 1.12 seconds to send and, leaving a gap for an acknowledgement of 320 ms, this makes the whole sequence 1.44 seconds long. In the fast mode at 200 baud each packet is 28 bytes long.

When a link is to be established the transmitting station sends data at 100 baud. This initial packet contains the callsign of the station being called. This is repeated at 200 baud. If the receiving station only receives the data sent at 100 baud, one type of acknowledgement is sent and communication is established at this speed. If the data sent at 200 baud is received correctly then another acknowledgement is sent, and the contact proceeds at 200 baud. Once established the contact then continues in the same way as a normal AMTOR contact.

2.2.10 Slow scan television

Slow scan television (SSTV) is used for a number of applications on the HF bands. For example, it is commonly used by radio amateurs to send pictures across the globe.

It uses many of the basic principles of ordinary broadcast television. However, broadcast television transmissions occupy bandwidths of several megahertz and obviously this cannot be accommodated on the HF bands. To overcome the problem, the rate at which the picture is transmitted is slowed down so that it can be accommodated within the 3 kHz bandwidth occupied by a single sideband transmission.

To create an SSTV signal a picture is scanned in basically the same way as a normal television signal. A point of light is moved across the picture and the reflected light level measured. Once the point has moved right across the picture it quickly moves back to the original side but down slightly and starts again. In this way the whole of the picture is scanned. At the receiving end the light levels detected at the transmitter are used to build up the original picture. The scanning directions are left to right for the line scan and top to bottom for the frame. Generally there are 120 or 128 lines per frame and the aspect ratio, i.e. the width to height ratio is 1:1.

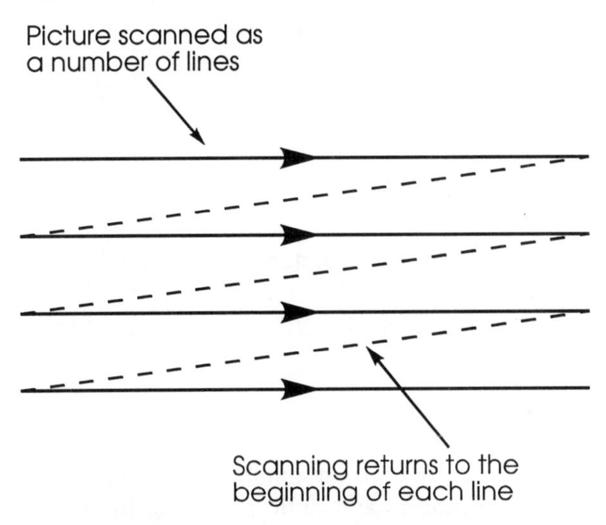

Figure 2.14 *Scanning a picture*

In order to ensure that the receiver and transmitter are synchronized, pulses are placed between each line and each complete picture or frame. Line synchronization pulses are 5 ms long whereas those for synchronising the whole picture or frame are 30 ms.

An example of a typical video signal is shown in Figure 2.15, and this is used to modulate the carrier. To achieve this an audio tone is frequency modulated and this is transmitted as a single sideband signal. The audio signal a frequency of 1200 Hz for a frame pulse, 1500 Hz for black and up to 2300 Hz which gives peak white. In essence this causes the radio-frequency signal to vary by +400 Hz for peak white, –400 Hz for black, and –700 Hz for a synchronization pulse. As the synchronization pulses are represented by a frequency lower than the one representing the black level, they are said to be blacker than black and they cannot be seen on the screen.

Picture quality can vary widely. To ensure that the optimum quality is achieved it is essential to have good detection of the synchronization pulses. Often a special filter is used to detect these 1200 Hz pulses, even though the pulses can easily be seen in the demodulated video signal.

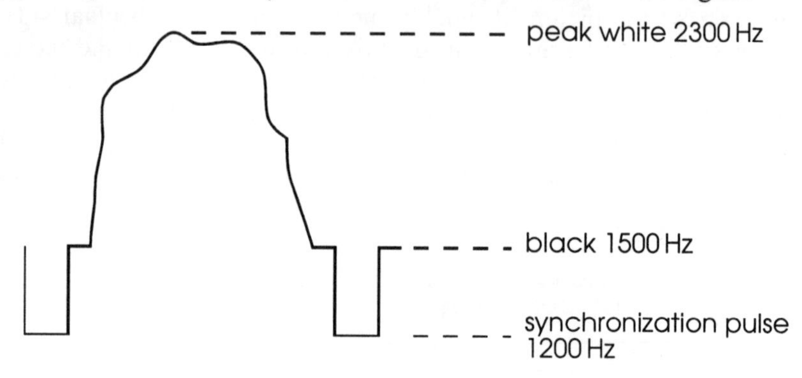

Figure 2.15 *A slow scan television video signal*

There are many standards for picture size. Typically pictures are 128 lines and take eight seconds to send. Another is 256 lines, although pictures can be almost any length, terminated by a frame synchronization pulse.

2.2.11 Facsimile

Fax or facsimile is required by a number of users including weather organizations, press agencies and radio amateurs. In essence the system used resembles that employed by slow scan television, but picture transmission times are much longer because of the higher definition which is required. Often a picture may take 15 minutes or longer. A further difference is that no synchronization pulses are used. Instead the accuracy of the transmitting and receiving systems is relied upon.

Originally pictures were wrapped around a drum, and a light sensitive cell was slowly moved along the picture as it rotated. The variations in light intensity were then used to modulate a carrier. At the receiver a similar rotating drum was used to reproduce the picture on specially sensitized paper. Today computers are often used, particularly at the receiving end, but the same transmission standard is employed.

Once the fax signal has been obtained, it is modulated on to a carrier for transmission. For the HF bands the carrier is frequency modulated and has an offset of +400 Hz for peak white and –400 Hz for black. Intermediate values of grey give frequencies between these two values.

A number of transmission formats are used. Figures quoted consist of two main parts, e.g. 120/288. The first of these figures is the number of lines sent per minute and the second is called the Index of Co-operation (IOC). The lines per minute is governed by the rotation speed of the scanning drum in the fax machine. For each line the drum has to complete one rotation. Standard speeds are 60, 90, 120 and 240 lines per minute. The IOC figure is a little more complicated, but basically defines the aspect ratio or width to height ratio of the picture. It is calculated by multiplying the drum diameter by the line density. The most common values for the IOC are 288 and 576.

2.3 Transmission codes

It is often useful to be able to define a type of transmission in an easy abbreviated form. The system which is currently in use was devised at the World Administrative Radio Conference in 1979 and used with effect from 1982. The code consists of three characters as shown in Table 2.5, at the end of this chapter. The first is a letter which describes the type of modulation. The second is a number which defines the nature of the modulating signal, that is, the information or signal which is being carried by the transmission. The third is another letter indicates the type of information being transmitted, such as Morse or audio, perhaps a speech transmission. For example, J3E denotes a single sideband suppressed carrier transmission and A1A is a Morse transmission.

In some instances the bandwidth of the transmission is also included after the basic code. This consists of a four-character code consisting of three numbers and a letter. A letter in the place of the decimal point denotes the units, namely H for hertz, k for kHz, M for MHz, and G for GHz. In this way a transmission occupying 12.5 kHz would be shown by 12k5.

Table 2.3 ASCII (American Standard Code for Information Interchange)

Hex	*Decimal*	*Keypress*	*Result*
00	00	CTRL @	NUL
01	01	CTRL A	SOH
02	02	CTRL B	STX
03	03	CTRL C	ETX
04	04	CTRL D	EOT
05	05	CTRL E	ENQ
06	06	CTRL F	ACK
07	07	CTRL G	BEL
08	08	CTRL H	BS
09	09	CTRL I	HT
0A	10	CTRL J	LF
0B	11	CTRL K	VT
0C	12	CTRL L	FF
0D	13	CTRL M	CR
0E	14	CTRL N	SO
0F	15	CTRL O	SI
10	16	CTRL P	DLE
11	17	CTRL Q	DC1 (X on)
12	18	CTRL R	DC2
13	19	CTRL S	DC3 (X off)
14	20	CTRL T	DC4
15	21	CTRL U	NAK
16	22	CTRL V	SYN
17	23	CTRL W	ETB
18	24	CTRL X	CAN
19	25	CTRL Y	EM
1A	26	CTRL Z	SUB
1B	27	CTRL [	ESC
1C	28	CTRL \	FS
1D	29	CTRL]	GS
1E	30	CTRL ^	RS
1F	31	CTRL -	US
20	32	space	SP
21	33	!	!
22	34	"	"
23	35	£	£
24	36	$	$
25	37	%	%
26	38	&	&
27	39	'	'
28	40	(	(

Hex	*Decimal*	*Keypress*	*Result*
29	41	)	)
2A	42	*	*
2B	43	+	+
2C	44	,	,
2D	45	-	-
2E	46	.	.
2F	47	/	/
30	48	0	0
31	49	1	1
32	50	2	2
33	51	3	3
34	52	4	4
35	53	5	5
36	54	6	6
37	55	7	7
38	56	8	8
39	57	9	9
3A	58	:	:
3B	59	;	;
3C	60	<	<
3D	61	=	=
3E	62	>	>
3F	63	?	?
40	64	@	@
41	65	A	A
42	66	B	B
43	67	C	C
44	68	D	D
45	69	E	E
46	70	F	F
47	71	G	G
48	72	H	H
49	73	I	I
4A	74	J	J
4B	75	K	K
4C	76	L	L
4D	77	M	M
4E	78	N	N
4F	79	O	O
50	80	P	P
51	81	Q	Q
52	82	R	R
53	83	S	S

<table>
<tr><th>Hex</th><th>Decimal</th><th>Keypress</th><th>Result</th></tr>
<tr><td>54</td><td>84</td><td>T</td><td>T</td></tr>
<tr><td>55</td><td>85</td><td>U</td><td>U</td></tr>
<tr><td>56</td><td>86</td><td>V</td><td>V</td></tr>
<tr><td>57</td><td>87</td><td>W</td><td>W</td></tr>
<tr><td>58</td><td>88</td><td>X</td><td>X</td></tr>
<tr><td>59</td><td>89</td><td>Y</td><td>Y</td></tr>
<tr><td>5A</td><td>90</td><td>Z</td><td>Z</td></tr>
<tr><td>5B</td><td>91</td><td>[</td><td>[</td></tr>
<tr><td>5C</td><td>92</td><td>\</td><td>\</td></tr>
<tr><td>5D</td><td>93</td><td>]</td><td>]</td></tr>
<tr><td>5E</td><td>94</td><td>^</td><td>^</td></tr>
<tr><td>5F</td><td>95</td><td>_</td><td>_</td></tr>
<tr><td>60</td><td>96</td><td>`</td><td>`</td></tr>
<tr><td>61</td><td>97</td><td>a</td><td>a</td></tr>
<tr><td>62</td><td>98</td><td>b</td><td>b</td></tr>
<tr><td>63</td><td>99</td><td>c</td><td>c</td></tr>
<tr><td>64</td><td>100</td><td>d</td><td>d</td></tr>
<tr><td>65</td><td>101</td><td>e</td><td>e</td></tr>
<tr><td>66</td><td>102</td><td>f</td><td>f</td></tr>
<tr><td>67</td><td>103</td><td>g</td><td>g</td></tr>
<tr><td>68</td><td>104</td><td>h</td><td>h</td></tr>
<tr><td>69</td><td>105</td><td>i</td><td>i</td></tr>
<tr><td>6A</td><td>106</td><td>j</td><td>j</td></tr>
<tr><td>6B</td><td>107</td><td>k</td><td>k</td></tr>
<tr><td>6C</td><td>108</td><td>l</td><td>l</td></tr>
<tr><td>6D</td><td>109</td><td>m</td><td>m</td></tr>
<tr><td>6E</td><td>110</td><td>n</td><td>n</td></tr>
<tr><td>6F</td><td>111</td><td>o</td><td>o</td></tr>
<tr><td>70</td><td>112</td><td>p</td><td>p</td></tr>
<tr><td>71</td><td>113</td><td>q</td><td>q</td></tr>
<tr><td>72</td><td>114</td><td>r</td><td>r</td></tr>
<tr><td>73</td><td>115</td><td>s</td><td>s</td></tr>
<tr><td>74</td><td>116</td><td>t</td><td>t</td></tr>
<tr><td>75</td><td>117</td><td>u</td><td>u</td></tr>
<tr><td>76</td><td>118</td><td>v</td><td>v</td></tr>
<tr><td>77</td><td>119</td><td>w</td><td>w</td></tr>
<tr><td>78</td><td>120</td><td>x</td><td>x</td></tr>
<tr><td>79</td><td>121</td><td>y</td><td>y</td></tr>
<tr><td>7A</td><td>122</td><td>z</td><td>z</td></tr>
<tr><td>7B</td><td>123</td><td>{</td><td>{</td></tr>
<tr><td>7C</td><td>124</td><td>|</td><td>|</td></tr>
<tr><td>7D</td><td>125</td><td>}</td><td>}</td></tr>
<tr><td>7E</td><td>126</td><td>~</td><td>~</td></tr>
<tr><td>7F</td><td>127</td><td>del</td><td>del</td></tr>
</table>

Table 2.4 Control Codes Names

NUL	Null (blank)
SOH	Start of Header
STX	Start of Text
ETX	End of Text
EOT	End of Transmission
ENQ	Enquiry
ACK	Acknowledgement
BEL	Bell (Audible)
BS	Backspace
HT	Horizontal Tab
LF	Linefeed
VT	Vertical Tab
FF	Form Feed
CR	Carriage Return
SO	Shift Out
SI	Shift In
DLE	Data Link Escape
DC1	Device Control 1
DC2	Device Control 2
DC3	Device Control 3
DC4	Device Control 4
NAK	Negative Acknowledgement
SYN	Synchronous Idle
ETB	End of Transmission Block
CAN	Cancel
EM	End of Medium
SUB	Substitute
ESC	Escape
FS	File Separator
GS	Group Separator
RS	Record Separator
US	Unit Separator
DEL	Delete

Table 2.5 Transmission Codes

First Character *Type of modulation*	Second Character *Nature of modulating signal*	Third Character *Information being transmitted*
N Emission of an unmodulated carrier	**0** No modulating signal	**N** No information transmitted
Amplitude Modulated Carriers	**1** A single channel containing digital or quantized information without the use of a subcarrier	**A** Telegraphy for aural reception
A Double sideband	**2** A single channel containing digital or quantized information with the use of a subcarrier	**B** Telegraphy for automatic reception
H Single sideband full carrier	**3** A single channel containing analogue information	**C** Facsimile
R Single sideband reduced level carrier	**7** Two or more channels containing digital or quantized information	**D** Data transmission
J Single sideband suppressed carrier	**8** Two or more channels containing analogue information	**E** Telephony
B Independent sideband	**9** Composite signal containing digital or quantized information together with one or more channels containing analogue information	**F** Television
C Vestigial sideband	**0** Any cases not covered	**W** Combination of the above
Angle Modulated Carrier		**X** Any cases not covered
F Frequency modulation		
G Phase modulation		
D Combination of angle and amplitude modulation		
Pulse Modulation		
P Unmodulated sequences of pulses		
K Pulses modulated in amplitude		
L Pulses modulated in width		
M Pulses modulated in position or phase		
Q Series of pulses in which the carrier is angle modulated in the period of the pulse		
V Combination of the above produced by other means		
W A combination of two or more of the following: angle, amplitude or pulse modulation		
X Any cases not covered		

3 Radio waves

A basic knowledge of radio waves and the way in which they travel is one of the key tools for the short wave listener. Knowing the best frequencies and times gives a listener a better chance of picking up a signal from a particular area. A simple example can be seen on the medium wave band where stations can be heard over relatively short distances during the day, but at night signals from further afield appear. Similarly signals in the short wave bands can be heard from many parts of the globe, although the time of day, the season and other factors affect the way the signals travel and therefore the distances which can be achieved.

The study of radio propagation, how radio waves travel, is fascinating, although trying to forecast exactly what radio conditions will be like is difficult. This is because of the large number of factors which affect it. Even so, short wave broadcast stations like the BBC World Service have whole departments devoted to determining the best frequencies for their transmissions. Fortunately for the short wave listener matters are not so critical, but by knowing how the bands are behaving it is possible to have a greater chance of hearing the types of station which are wanted.

3.1 Radio waves

Radio signals are electromagnetic waves. They are the same type of radiation as light, ultraviolet and infrared rays, differing from them in their wavelength and frequency.

These waves are quite complicated, having both electric and magnetic components. Many of their properties, however, can be visualized by comparing them to a wave on the sea, or a pond.

There are a number of properties of a wave. The first is its **wavelength**. This is the distance between a point on one wave to the corresponding point on the next as shown in Figure 3.1. One of the most obvious points to choose is the peak as this can be easily identified.

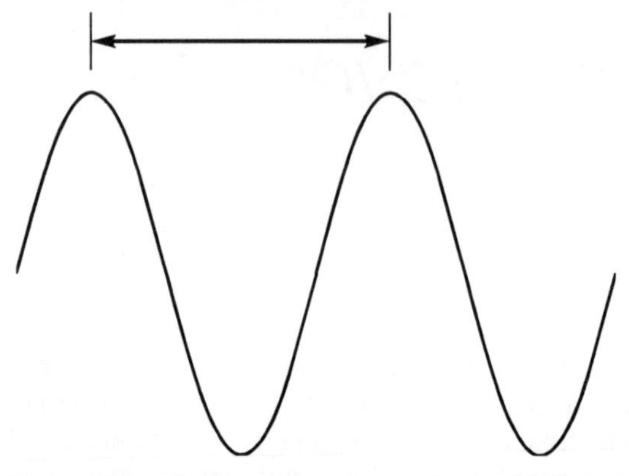

Figure 3.1 *The wavelength of an electromagnetic wave*

The second is its **frequency**. Returning to the sea or pond analogy, this is the number of times a particular point on the water surface moves up and down in a given time, normally a second. The unit of frequency is the hertz as mentioned in Chapter 2. This unit is named after the German scientist who discovered radio waves.

The third major property of the wave is its **velocity**. Radio waves travel at the same speed as light. For most practical purposes this speed is taken as 300 000 000 metres per second. However, a more exact value is 299 792 500 metres per second.

Many years ago the position of stations on the radio dial were given as wavelengths. A station in the medium wave band might have a wavelength of 247 metres. Today stations give their frequency because this is easier to determine. A simple instrument like a frequency counter can measure this very accurately. It is simple to relate these two parameters as they are linked by the speed of the wave, the same as the speed of light or 300 000 000 as shown:

$$\lambda = \frac{c}{f}$$

where λ = the wavelength in metres, f = frequency in hertz and c = speed of radio waves (light) taken as 300 000 000 metres per second for all practical purposes.

Taking the previous example, the wavelength 247 metres corresponds to a frequency of 300 000 000/247 or 1.2145 million hertz or megahertz.

3.2 Radio spectrum

The radio spectrum covers a vast range. At the bottom end of the spectrum there are signals of a few kHz, whereas at the top end new semiconductor devices are being developed which operate at frequencies of a hundred GHz and more. Between these extremes lie the frequencies with which we are familiar, still a huge spread of spectrum space available for transmissions. To make it easy to refer to different portions of the spectrum, designations are given to them as shown in Figure 3.2.

It can be seen from this figure that transmissions in the long wave broadcast band between 140.5 and 283.5 kHz fall into the low frequency or LF portion of the spectrum. Other types of transmission which are made on low frequencies include navigational beacons which transmit on frequencies around 100 kHz and sometimes less.

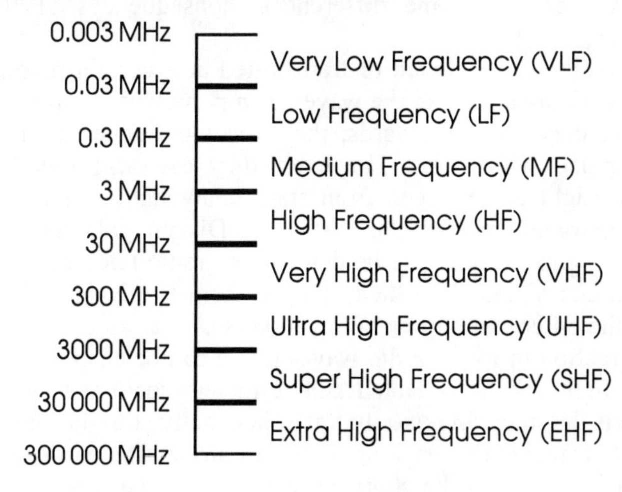

Figure 3.2 *The frequency spectrum*

Moving up in frequency, the medium wave broadcast band falls into the medium frequency or MF portion of the spectrum. Above this broadcast band is where the lowest frequency short wave bands are sometimes considered to start. Here there is an amateur radio band together with allocations for maritime communications.

Between 3 and 30 MHz is the high frequency or HF portion. Within this frequency range lie the true short wave bands. Signals from all over the world can be heard. Broadcasters, radio amateurs and many others use them.

Moving up in frequency, the very high frequency or VHF part of the spectrum contains a large number of mobile users. 'Radio Taxis' and the emergency services have allocations here, as do the familiar VHF FM broadcasts.

In the ultra high frequency or UHF part of the spectrum most of the terrestrial television stations are located. In addition to these there are more mobile users including the increasingly popular cellular telephones.

Above this in the super high frequency or SHF and extremely high frequency or EHF portions of the spectrum there are many uses for the radio spectrum. These are used increasingly for commercial satellite and point to point communications.

3.3 How waves travel

Radio signals are similar to light waves and behave in a similar way. Obviously there are some differences, consequences of their lower frequencies.

When a signal is radiated or transmitted at a certain point, the radio waves travel outwards, like the waves on a pond when a stone is dropped into it. As they move outwards, they become weaker as they have to cover a much wider area. However, they can still travel enormous distances. Light can be seen from stars many light years away. Radio waves can travel over similar distances. Distant galaxies and quasars emit radio signals. These can be detected by radio telescopes which pick up the minute signals and then analyse them to give us clues to what exists in the outer extremities of the universe.

The direction in which radio waves travel can be changed. In the same way that light can be reflected and refracted, so too can radio waves. This means that they do not only travel in exactly straight lines. They can be made to travel over very long distances around the earth. It is because of this that signals on the short wave bands can be heard coming from places all over the world.

3.4 Polarization

Light waves can be **polarized**. Special polaroid lenses allow only light of a particular polarization through. This can be very useful on sunny days when reflections from surfaces can cause glare. As the reflected light has a particular polarization, it is possible to reduce the glare by using polaroid lenses to cut out the reflections.

Radio waves are also polarized. As electromagnetic waves consist of electric and magnetic components in different planes, the plane of the electric wave is taken. The polarization of a radio wave can be very important because aerials are sensitive to polarization and generally only pick up or transmit signals having a particular polarization.

3.5 Ground wave

Before we look at how radio waves travel all round the world, let us first look at the way in which a signal travels away from the aerial. When a signal is transmitted from an aerial it spreads out and can be picked up by receivers which are in the line of sight. On frequencies in the long and medium wave bands particularly, stations can be received over greater distances than this. This happens because the signals tend to follow the earth's curvature. This is called the **ground wave**. It occurs because currents are induced in the earth's surface. This slows the wave front down nearest to the earth causing it to tilt downwards, enabling it to follow the curvature and travel over distances which are well beyond the horizon.

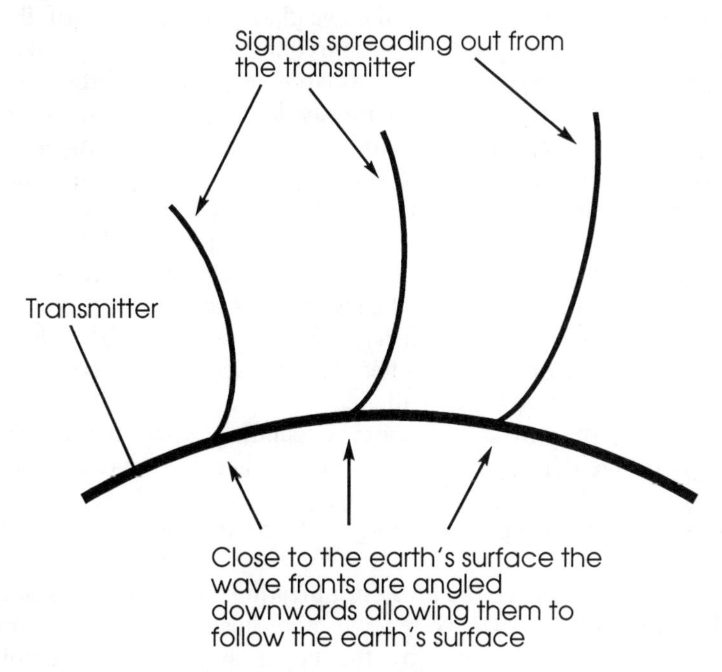

Figure 3.3 *A ground wave*

The ground wave can only be used for signals below about 2 MHz. As the frequency increases, the attenuation of the whole signal increases and the coverage is considerably reduced. Obviously the exact range will depend on many factors. Typically a high power medium wave station may be expected to be heard over distances of 100 miles. There are also many low power broadcast stations running a hundred watts or so. These might have a coverage area extending out 15 to 20 miles.

As the effects of attenuation increase with frequency, even very high power short wave stations are only heard over relatively short distances using ground wave. Instead these stations use reflections from layers high up in the atmosphere to cover greater distances.

3.6 Layers above the earth

The atmosphere above the earth consists of many layers as shown in Figure 3.4. Some of these have a considerable effect on radio waves. Closest to the surface is the **troposphere**. This region has little effect on short wave frequencies below 30 MHz, although at higher frequencies it plays a major role. At certain times transmission distances may be increased from a few tens of kilometres to a few hundred kilometres. This is the area which governs the weather, and in view of this the weather and radio propagation at these frequencies are closely linked.

Above the troposphere is the **stratosphere**. This has little effect on radio waves, but above it in the **mesosphere** and **thermosphere** the levels of ionization rise in what is collectively called the **ionosphere**.

In this area radiation from the sun, mainly in the form of ultraviolet light, causes some of the air molecules to ionize, forming free electrons and positively charged ions. As the air in these areas is relatively sparse it takes some time for them to recombine. These free electrons affect radio waves, causing them to bend back towards the earth.

The level of radiation starts to rise above altitudes of 30 km, but there are areas where the density is higher. These layers have been designated the letters D, E and F to identify them.

The degree of ionization varies, depending upon the amount of radiation received from the sun. At night when the layers are hidden from the sun, the level of ionization falls. The D layer virtually disappears and the E layer is greatly reduced in strength, often all but disappearing.

Other factors influence the level of ionization. One is the season of the year. In the same way that we receive more heat in summer, so too the amount of radiation received by the upper atmosphere is increased. Similarly the amount of radiation received in winter is less.

The number of sunspots on the sun has a major affect on the ionosphere. These spots indicate areas of high magnetic fields. The number of spots varies considerably. They have been monitored for over 200 years and it has been found that the number varies over a cycle of approximately 11 years. This figure is an average, and any particular cycle may vary in length from about nine to 13 years. At the peak of the cycle there may be as many as 200 spots whilst at the minimum the number may be in single figures and on occasions none have been detected.

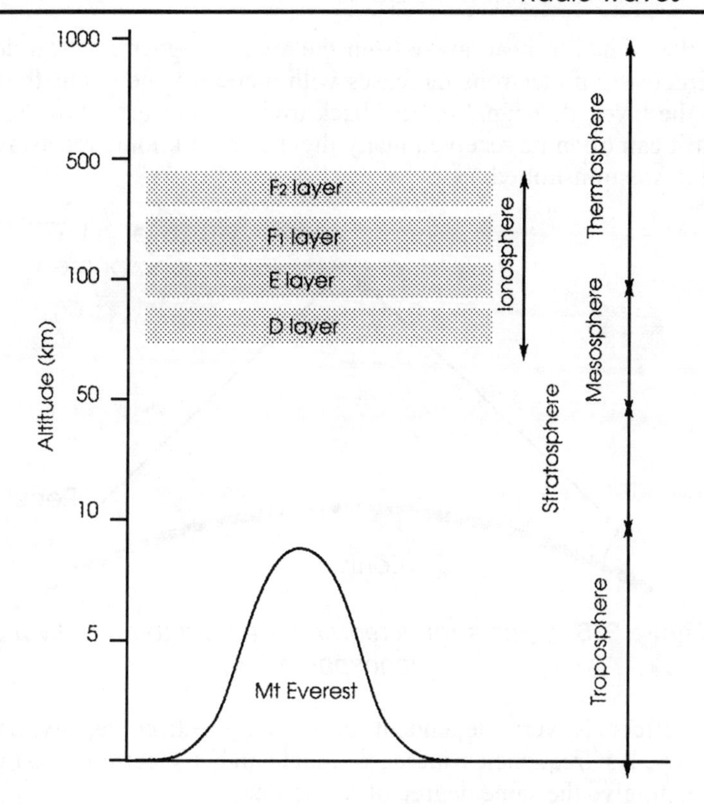

Figure 3.4 *Areas of the atmosphere*

Caution! Under no circumstances should you view the sun directly, or even through dark sunglasses. This is very dangerous and people have lost the sight of an eye trying it.

Sunspots affect radio propagation because they emit vast amounts of radiation. This increases the level of ionization in the ionosphere. Accordingly radio propagation varies in line with the sunspot cycle. The last peak occurred in 1991 and the next one is predicted to occur in 2002.

3.6.1 Ionization and radio waves

An exact explanation of the way in which the ionization in the atmosphere affects radio waves would be very complicated. However it is possible to gain an understanding of the basic concepts from a simpler explanation.

When the signal reaches the ionization, it sets the free electrons in motion and they act as if they formed millions of minute aerials. The electrons retransmit the signal, but with a slightly different phase. This causes the signal to bend away from the area of higher electron density. As the density of electrons increases with increasing height as the signal enters the layer, the signal is bent back towards the surface of the earth, so that it can often be received many thousands of kilometres away from where it was transmitted.

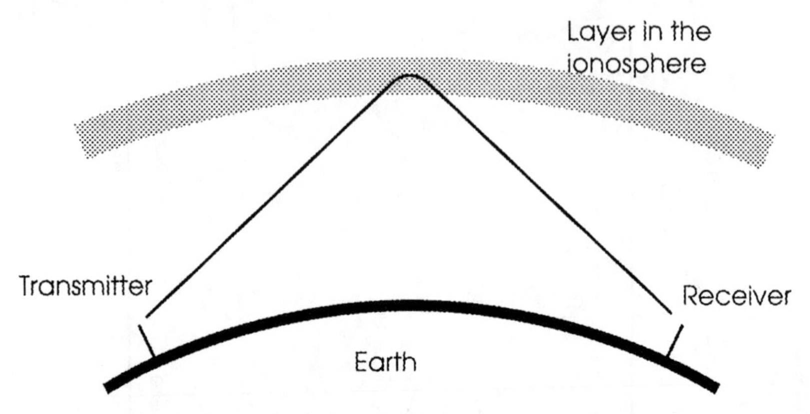

Figure 3.5 *Signals reflected and returned to earth by the ionosphere*

The effect is very dependent upon the electron density and the frequency. As frequencies increase, much higher electron densities are required to give the same degree of refraction.

3.6.2 Layers of ionization

Each of the bands or layers in the ionosphere acts in a slightly different way, affecting different frequencies.

The lowest layer is the D layer at a height of around 75 km. Instead of reflecting signals, this layer tends to absorb any signals which it affects. The reason for this is that the air density is very much greater at this altitude and power is absorbed when the electrons are excited. However, this layer only affects signals up to about 2 MHz or so. It is for this reason that only local ground wave signals are heard on the medium wave broadcast band during the day.

This layer has a relatively low electron density and levels of ionization fall relatively quickly. As a result the D layer is only present when radiation is being received from the sun and at night it is not present. When this happens, it means that low frequency signals can be reflected by higher layers. This is why signals from much further afield can be heard on the medium wave band at night.

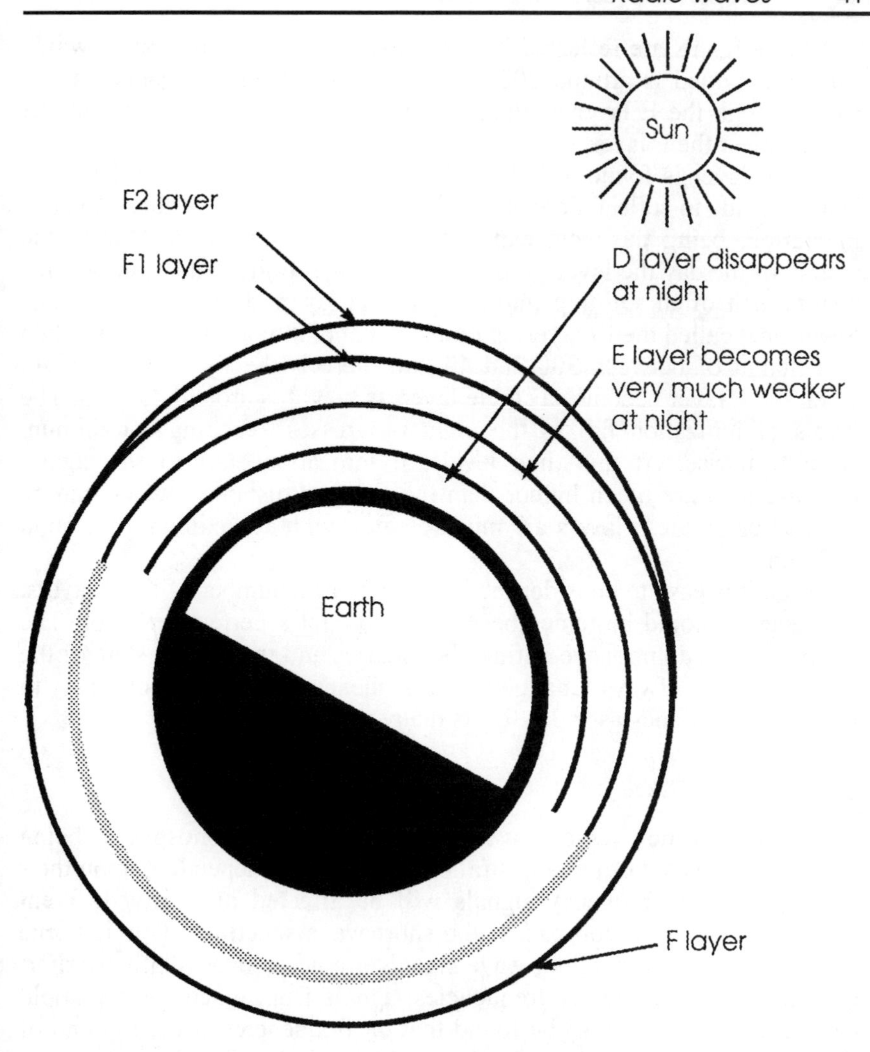

Figure 3.6 *Variation of the ionized layers during the day*

Above the D layer is the E layer. At a height of around 110 km, this layer has a higher level of ionization than the D layer. It also reflects many of the signals which reach it, rather than absorbing them. However, there is a degree of attenuation with any signal reflected by the ionosphere. The atmosphere is still relatively dense at the altitude of the E layer. This means that the ions recombine quite quickly and levels of ionization sufficient to reflect radio waves are only present during the hours of daylight. After sunset the number of free ions falls relatively quickly to a level where they usually have little effect on radio waves.

When signals are reflected by the E layer, the maximum range which can be obtained is around 2000 km. As in the D layer, signals start to pass through the E layer as the frequency rises. When this happens the signals meet the F layer.

The F layer is found at heights between 200 and 400 km. Like the E layer it tends to reflect signals which reach it. It has the highest level of ionization, being the most exposed to the sun's radiation. During the course of the day the level of ionization changes quite significantly. This results in it often splitting into two distinct layers during daytime. The lower one, called the F_1 layer, is found at a height of around 200 km, then at a height of between 300 and 400 km there is the F_2 layer. At night when the F layer becomes a single layer its height is around 250 km. The levels of ionization fall as the night progresses, reaching a minimum around sunrise. At this time levels of ionization start to rise again. Because they are much higher than the E layer, distances which can be reached using the F layers are much greater, with a maximum of around 4000 km.

Often it is easy to consider the ionosphere as a number of fixed layers. However it should be remembered that it is not a perfect reflector. The various layers do not have defined boundaries and the overall state of the ionosphere is always changing. This means that it is not easy to formulate hard-and-fast rules for its many attributes.

3.6.3 Skywaves

The way in which radio waves travel through the ionosphere, being absorbed, reflected or passing straight through it, is dependent upon their frequency. Low frequency signals will be affected in totally different ways to those at the top end of the short wave spectrum. This is borne out by the fact that medium wave signals are heard over relatively short distances, while at higher frequencies signals from much further afield can be heard. It may also be found that on frequencies at the top end of the short wave spectrum no signals may be heard on some days.

To explain how the effects change with frequency, consider a low frequency signal transmitting in the medium waveband at a frequency of f_1. The signal spreads out in all directions along the earth's surface as a ground wave which is picked up over the service area. Some radiation also travels up to the ionosphere. However because of the frequency in use the D layer absorbs the signal during daytime. At night the D layer disappears and the signal can then pass on, reflected by the higher layers.

Signals higher in frequency at f_2 pass straight through the D layer. When they reach the E layer they can be affected by it, being reflected back to earth. The frequency at which signals start to penetrate the D layer in the day is difficult to define as it depends on many factors. However it is often in the region of 2 to 3 MHz.

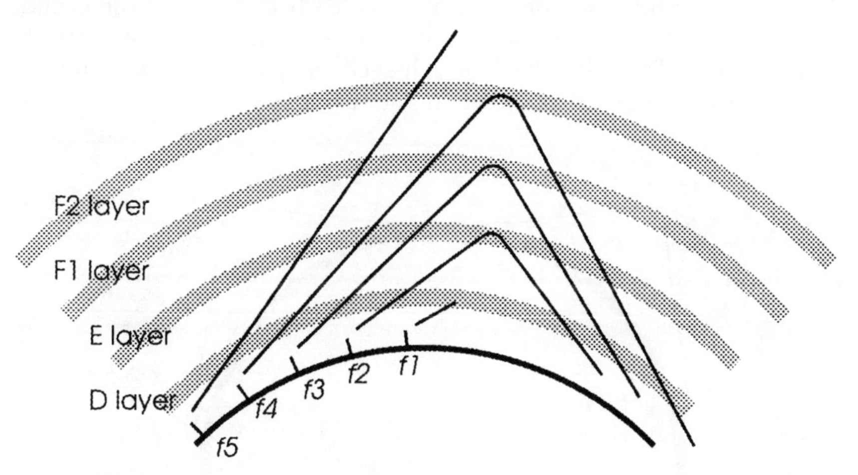

Figure 3.7 *Radio wave propagation at different frequencies*

Also as the frequency increases, the ground wave coverage decreases. Medium wave broadcast stations can be heard over distances of many tens of miles. At frequencies in the short wave bands this is much smaller. Above 10 MHz signals may be heard over only a few miles, depending upon the power and aerials being used.

The E layer tends to reflect only signals in the lower part of the short wave spectrum. As the frequency increases, signals penetrate further into the layer, eventually passing right through it. Once through, they travel on to the F layer. This may have split into the F_1 and F_2 layers. When the signals at a frequency of f_3 reach the first of the layers they are again reflected back to earth. Then as the frequency rises to f_4 they pass on to the F_2 layer where they are reflected. As the frequency rises still further to f_5 the signals pass straight through all the layers, travelling on into outer space.

During the day at the peak in the sunspot cycle it is possible for signals as high as 50 MHz and more to be reflected by the ionosphere. However this figure falls to below 20 MHz at very low points in the cycle.

To achieve the longest distances it is best to use the highest layers. This is achieved by using a frequency which is high enough to pass through the lower layers. From this it can be seen that frequencies higher in the short wave spectrum tend to travel the longest distances. Even so, it is still possible to hear signals from the other side of the globe on low frequencies at the right time of day, and if good aerials and high powers are used.

3.6.4 Multiple reflections

The maximum distance for a signal reflected from the F_2 layer is about 4000 km. However radio waves travel much greater distances than this around the world. This cannot be achieved using a single reflection, but instead several are used as shown in Figure 3.8.

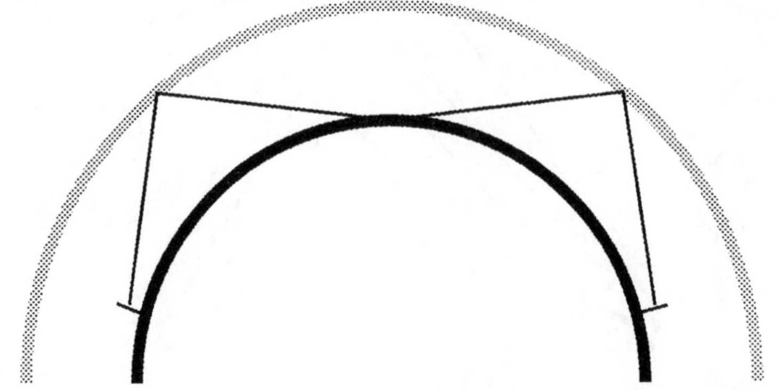

Figure 3.8 *Several reflections used to give greater distances*

To achieve this the signals travel to the ionosphere and are reflected back to earth in the normal way. Here they can be picked up by a receiver. However the earth also acts as a reflector because it is conductive. The areas which are more conductive act as better reflectors. The sea acts as an excellent reflector.

Once reflected at the earth's surface the signals travel towards the ionosphere where they are again reflected back to earth.

At each reflection the signal suffers some attenuation. This means that it is best to use a path which gives the minimum number of reflections as shown in Figure 3.9. Lower frequencies are more likely to use the E layer and as the maximum distance for each reflection is less, it is likely to give lower signal strengths than a higher frequency using the F layer to give less reflections.

3.6.5 Chordal hop

Not all reflections around the world occur in exactly the ways described. It is possible to calculate the path which would be taken, the number of reflections, and hence the path loss and signal strength expected. Sometimes signal strengths appear higher than would be expected. In conditions like these it is likely that a propagation mode called chordal hop is being experienced.

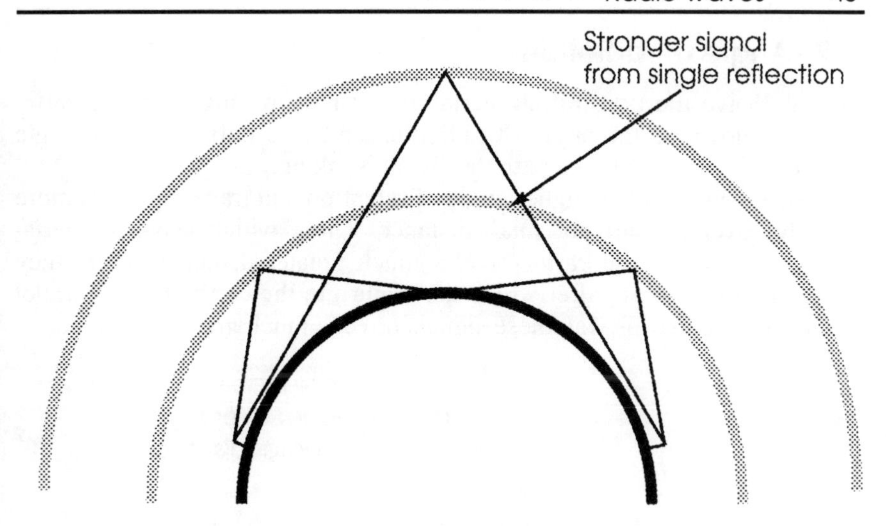

Figure 3.9 *The minimum number of reflections usually gives the best signal*

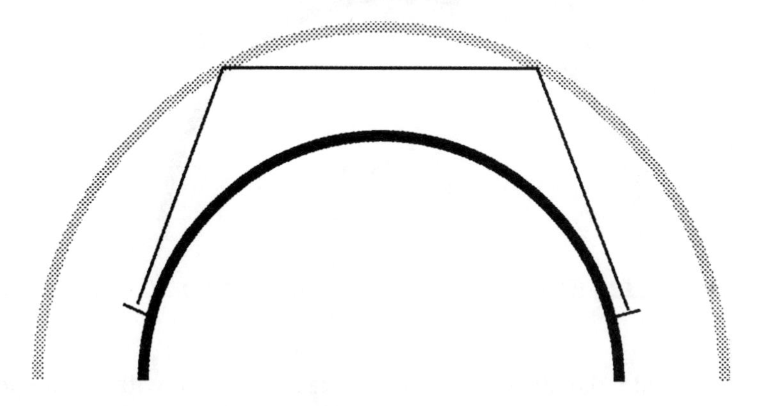

Figure 3.10 *Chordal hop propagation*

When this happens, the signal travels to the ionosphere where it is reflected, but instead of returning to the earth it takes a path which intersects with the ionosphere again, only then being reflected back to earth.

Because of the way in which this occurs, fewer reflections are needed to cover a given distance. As a result signal strengths are higher when this mode of propagation is used.

3.7 Angle of radiation

Signals leave the transmitting aerial at a variety of angles to the earth. This is known as the angle of radiation, and it is defined as the angle between the earth, and the path the signal is taking.

Those which have a higher angle of radiation and travel upwards more steeply cover a relatively small distance. Those which leave the aerial almost parallel to the earth travel a much greater distance before they reach the ionosphere, after which they return to the earth almost parallel to the surface. In this way these signals travel a much greater distance.

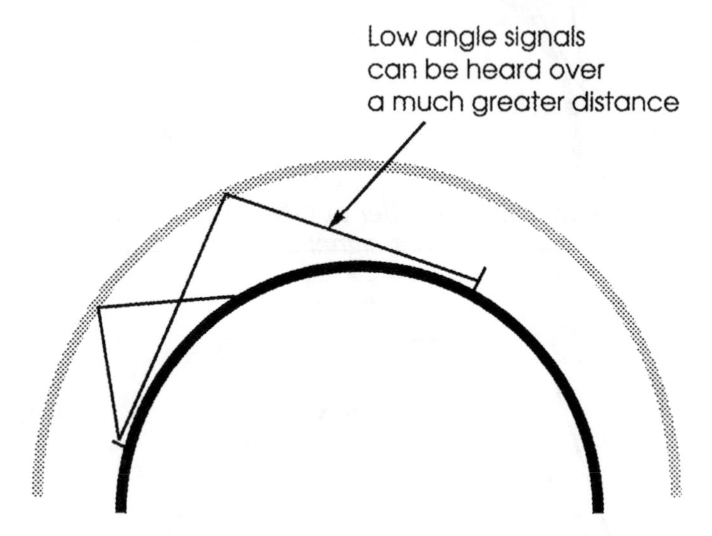

Figure 3.11 *Effect of the angle of radiation on the distance achieved*

To illustrate the difference this makes, a signal with an angle of radiation of 0 degrees can cover a distance of up to 2000 km for a reflection from the E layer and about 4000 km for the F layer. If the angle of radiation is increased to 20 degrees the distances fall to 400 km and 1000 km respectively.

For signals to travel the maximum distance it is imperative to have a low angle of radiation. However, broadcast stations often need to make their aerials directive to ensure the signal reaches the desired area. Not only do they ensure they are radiated with the correct azimuth, but they also ensure they have the correct angle of elevation or radiation so that they are beamed to correct area. This is achieved by altering the aerial parameters.

3.8 Critical frequency and critical angle

When a signal reaches a layer in the ionosphere it undergoes refraction and often it will be reflected back to earth. The steeper the angle at which the signal hits the layer the greater the degree of refraction is required. If a signal is sent directly upwards this is known as vertical incidence as shown in Figure 3.12.

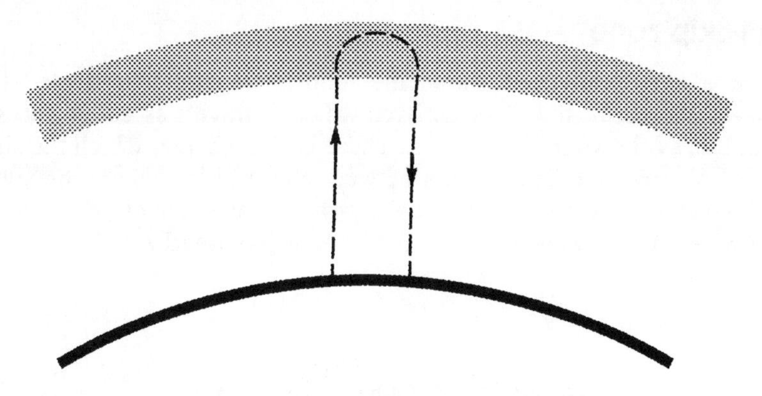

Figure 3.12 *Vertical incidence*

For vertical incidence there is a maximum frequency for which the signals will be returned to earth. This frequency is known as the **critical frequency**. Any frequencies higher than this will penetrate the layer and pass right through it on to the next layer or into outer space.

3.9 MUF

When a signal is transmitted over a long distance path, it penetrates further into the reflecting layer as the frequency increases. Eventually it passes straight through. This means that for a given path there is a maximum frequency which can be used. This is known as the **maximum usable frequency** or MUF. Generally the MUF is three to five times the critical frequency, depending upon which layer is being used and the angle of incidence.

For optimum operation a frequency about 20% below the MUF is generally used. The MUF also varies greatly depending upon the state of the ionosphere. Accordingly it changes with the time of day, season, position in the 11-year sunspot cycle and the general state of the ionosphere.

3.10 LUF

When the frequency of a signal is reduced, further reflections are required and losses increase. As a result there is a frequency below which the signal cannot be used. This is known as the **lowest usable frequency** or LUF.

3.11 Skip zone

When a signal travels towards the ionosphere and is reflected back towards the earth, the distance over which it travels is called the **skip distance,** as shown in Figure 3.13. There is an area over which the signal cannot be received. This occurs between the position where the signals start to return to earth and where the ground wave cannot be heard. The area where no signal is heard is called the **skip** or **dead zone**.

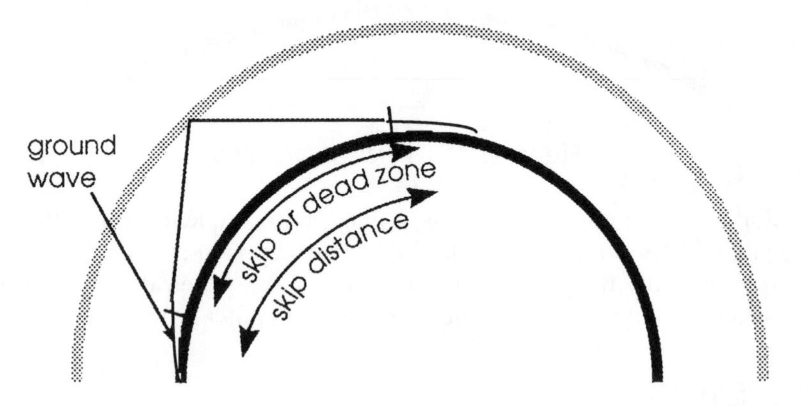

Figure 3.13 *Skip zone and skip distance*

3.12 State of the ionosphere

Radio propagation conditions are of great importance to a vast number of users of the short wave bands. Broadcasters, for example, are very interested in them as are other professional users. To detect the state of the ionosphere, an instrument called an **ionosonde** is used. This is basically a form of radar system which transmits pulses of energy up into the ionosphere. The reflections are then monitored and from them the height of the various layers can be judged. Also by varying the frequency of the pulses, the critical frequencies of the various layers can be judged.

3.13 Fading

One of the characteristics of listening to short wave stations is that some signals appear to fade in and out all the time. These alterations are taken as a matter of course by listeners who are generally very tolerant to the imperfections in the quality of the signal received from the ionosphere. There are a number of causes for fading but they all depend upon the fact that the state of the ionosphere is constantly changing.

The most common cause of fading is multipath interference. This occurs because the signal leaves the aerial at a variety of different angles and reaches the ionosphere over a wide area. As the ionosphere is very irregular the signal takes a number of different paths as shown in Figure 3.14. The changes in the ionosphere cause the lengths of these different paths to vary. This means that when the signals come together at the receiving aerial they pass in and out of phase with one another. Sometimes they reinforce one another and at other times they cancel each other out. This results in the signal level changing significantly over periods of even a few minutes.

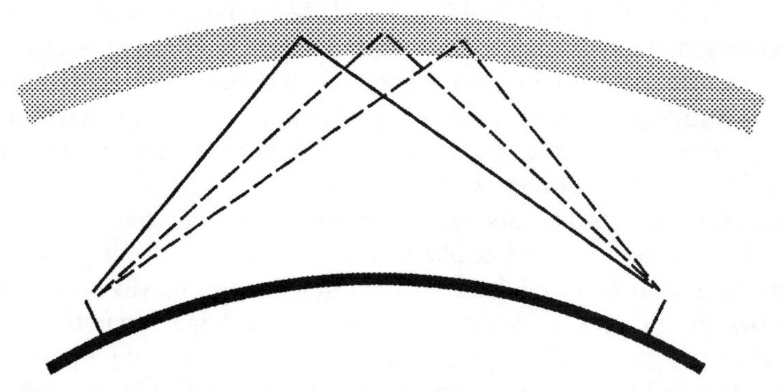

Figure 3.14 *Signals can reach the receiver via several paths*

Another reason for signal fading is changes in polarization. When the ionosphere reflects signals back to earth they may have any polarization. For the best reception, signals should have the same polarization as the receiving aerial. As the polarization of the reflected wave will change dependent upon the ionosphere, the signal strength will vary according to the variations in polarization.

If the receiver is on the edge of the skip zone for a particular signal, any slight variations in the state of the ionosphere will cause the receiver to pass into or out of the skip zone, giving rise to signal strength variations.

Sometimes severe distortion can be heard, particularly on amplitude modulated signals. This occurs when the various sideband frequencies are affected differently by the ionosphere. This is called **selective fading** and it is often heard most distinctly when signals from the ground and skywaves received together.

3.14 Beacons

A vast amount of research has been undertaken concerning the ionosphere so that predictions can be made more accurate. This has improved the prediction of overall trends, but the day-to-day forecasts of conditions are still difficult to make with any accuracy. In many ways it is like trying to forecast the weather accurately! For listeners to be able to gain an idea of which areas can be heard on any given day and time there is no substitute for actually listening and finding out what is on. To help give a quick idea of the regions to which a band may be open **beacon stations** are used.

The International Telecommunications Union (ITU) has instigated the installation of a number of beacons as part of a study of HF propagation. A number of stations are set up around the world, using the frequencies 5470, 7870, 10 407, 14 405 and 20 945 kHz. The stations step through the frequencies, spending four minutes on each, returning to the first after 20 minutes. Different stations have different time slots for each frequency and as a result no clashes occur. By multiplexing the frequencies in this way listeners can gain a good idea of conditions without having to monitor a vast different number of frequencies for the different stations.

Amateur radio enthusiasts have also set up beacon chains around the world. The stations are generally low powered, transmitting an almost continuous signal with regular identification, usually in Morse. Currently a number of stations transmitting a continuous signal operate between 28.2 and 28.3 MHz although a time multiplexed system operates on 14.100 MHz. Other systems are planned for 21.150 MHz and it is expected that beacons between 28.2 and 28.3 MHz will be rationalized to follow this type of scheme.

3.15 Sporadic E

Sometimes in summer it is possible for signals to be audible in the bands at the top end of the short wave spectrum at the bottom of the sunspot cycle. When the maximum usable frequency may be well below the frequencies in question, signals from stations up to 2000 km distance may be heard in summer. This occurs as a result of a form of propagation known as **sporadic E**.

Sporadic E used to be well known when television used frequencies around 50 MHz. Sometimes in summer reception would be disturbed by interference from distant stations. Even today reception of VHF FM signals can be disturbed when frequencies around 100 MHz are affected by it. The maximum frequencies generally affected are around 150 MHz, although the maximum ever recorded was 215 MHz. When it occurs at these high frequencies amateurs using the 144 MHz band make good use of it even though it may last for only a few minutes.

Sporadic E occurs as a result of highly ionized areas or clouds forming within the E layer. These clouds may measure only about 100 km across, and this means that propagation is quite selective with signals coming from a particular area.

The ionization slowly builds up and at first only low frequencies are affected. As the levels of ionization increase frequencies above those normally supported by the E layer are reflected.

Sporadic E normally occurs in the summer, reaching a peak broadly around the longest day. Even so openings at the top of the short wave band may be found at least a couple of months either side of this. The openings for frequencies well into the VHF portion of the spectrum are normally confined to the centre of the season because high ionization levels are required. Moreover the openings for these very high frequencies do not last as long as those for the lower ones because of the steady build up and decay.

The sporadic nature of this form of propagation means that it is difficult to predict when it will occur. Even when propagation is supported by this mode, it is very variable, particularly at the very high frequencies. The ionized clouds are blown about in the upper atmosphere by the swiftly moving air currents. This means that the area from which stations are heard can change.

3.16 Ionospheric disturbances

At certain times ionospheric propagation can be disrupted and signals on the short wave bands can disappear completely. There are two main types of disturbance, but both are caused by solar activity. This often reaches a maximum at the peak of the sunspot cycle. Although radio propagation conditions are better, they are also more susceptible to being disturbed by these ionospheric disturbances.

The first type of disturbance is called a **sudden ionospheric disturbance** or SID. When a SID occurs, signals on the short wave bands disappear for anything from a few minutes to a few hours. These disturbances are caused by solar flares on the surface of the sun which send out vast quantities of radiation. When this radiation reaches the earth it increases the level of ionization in the D layer, causing absorption of signals over a large part of the short wave spectrum. As the radiation is received directly from the sun, these disturbances only occur on the daylight side of the earth.

A more devastating form of disturbance is a complete fade out or ionospheric storm. When large flares occur, vast quantities of ionized material leave the sun. This takes up to 48 hours to reach the earth. When it arrives it overloads the earth's protective mechanisms and causes massive increases in the level of ionization. This also causes the absorption in the D layer to increase, wiping out signals over most of the short wave spectrum. Unlike the SID, the effects of a full ionospheric storm may last for a few days, with the bands only slowly returning to normal.

3.17 Propagation on different frequencies

After spending some time listening on the short wave bands, it is possible to get a 'feel' for the types of stations which will be heard on different frequencies at different times. This knowledge of the bands is an essential tool for the listener. It will enable him or her to choose the right frequency to pick up a station from a particular area.

It is not possible to give an exact explanation of what might be heard for each band of frequencies, but a broad outline of what to expect can be given. The chapters on amateur radio and broadcasting give more detailed explanations of the relevant amateur and broadcast bands.

To generalize, the frequencies up to 3 or 4 MHz will support relatively local communications. At the lower frequencies signals may be heard over distances of 30 or 40 miles dependent upon the powers and aerials. Around 4 MHz signals over a hundred miles may be heard during the day, with large increases at night.

As the frequency increases so does the distance over which stations can be heard. Around 15 MHz it is possible to hear stations over distances of 2500 km or more during the day. Often stations much further afield can be heard, especially towards dusk and after nightfall. At these frequencies the bands may close during the night, especially in winter or during the low parts of the sunspot cycle.

As the frequency increases still further, daylight propagation distances increase, although the bands become more dependent upon the 11-year sunspot cycle. At frequencies around 30 MHz it is unlikely that any signals will be heard, except by ground wave, in the years either side of the sunspot minimum, except by sporadic E in the summer. At the peak, however, these frequencies support global communications with ease during the day, but usually closing soon after dark.

3.18 Propagation prediction

Whilst it is possible to gain a feel for how the various areas of the radio spectrum behave, predictions for the radio conditions are published in some publications, or broadcast over the air. These predictions can be very valuable for the listener, enabling him or her to know when to listen.

It is not possible to predict exactly what the conditions will be like. Instead they are an indication of what is expected and they should only be treated as a probability, giving the chances that a particular path or route on a frequency will be open for radio communications.

The predictions are also dependent upon the aerial in use and the exact location. Most predictions indicating a particular path will relate to the capital of the country. In other words any predictions for the UK will normally relate to a station located in London. Also the aerial in use will have a marked effect. A poor aerial will reduce the chance of hearing stations from a given area. However with a little experience it is possible to read the predictions and adapt them for one's own location and set up.

Propagation predictions can be found in a number of places. Amateur radio magazines often carry them in a special section. These are often in the form of a probability that stations from given area can be heard.

A number of transmitting stations carry predictions. Of these the most famous is WWV, an American standard frequency transmission found on 2.500, 5.000, 10.000 and 20.000 MHz. The station is primarily used as a highly accurate frequency transmission, but it also transmits bulletins giving the solar flux, A index for the previous day, the Boulder K index, state of solar activity and the condition of the earth's magnetic field.

The solar flux is a measure of the sun's radiation at a frequency of 2800 MHz. A figure of 60 indicates that conditions will be quiet, but as it rises it gives an indication of an improvement. Readings of 80 and higher mean that the higher frequency bands above 20 MHz or so will be open for signals. Flux readings around 200 occur towards the peak of the cycle and give an indication that even frequencies above 30 MHz can support long distance communication via the F_2 layer.

The A and K indices give a measure of the earth's magnetic field. Low values of these figures give an indication of better HF conditions. An A index of 10 or less is indicative of good HF conditions. The figure can rise to over 100, but this is not common.

The K index is similar to the A index, but is more up to date. A figure of between 0 and 1 indicates quiet geomagnetic field conditions and the likelihood of good HF conditions. 1 to 3 indicates unsettled conditions and 4 and above show high levels of geomagnetic activity and poor HF conditions.

4 Receivers

The radio receiver is the central piece of equipment in any short wave listening station. Its performance and the way in which it is used govern the results that can be achieved. Receivers today use much of the latest technology and have a number of controls not found on normal broadcast sets. Because most of the short wave bands have an enormous number of signals of different types, it may take a little while to become familiar with the best ways to use the set to separate the required station from all the other signals. This is a skill which is learned over a period of time. Once gained it enables far more to be achieved with the set. Often a skilled operator will be able to use a receiver to pull signals out of the noise when a newcomer might not even realize they were there.

4.1 Receiver controls and connections

There is a vast choice of types of receiver on the market. Despite this there is a degree of commonality in the controls which are used. This means that it is usually quite easy to swap from one receiver to another. Obviously the more specialized functions like scanning will vary to a greater degree, but even with these experienced listeners can usually use them without reference to the operating manual.

4.1.1 Tuning

The first control which is noticed on almost any receiver is the tuning control. For analogue sets which have a dial and a variable capacitor this presents very little difficulty, although tuning in some signals like SSB can take a little practice. More modern sets often use frequency synthesizers. With these it is often possible to enter the frequency from a keypad as well as using a tuning knob in the traditional way. This gives

Figure 4.1 *A communications receiver (Courtesy of Lowe Electronics)*

the advantage that it is possible to set the frequency of a known station very quickly or move to another band without having to turn the knob many times.

Some synthesized sets also have an option to change the step size or tuning speed. This can be done because a synthesized receiver changes its frequency in distinct steps. Normally these steps are small enough not to be noticed, but sometimes it is an advantage to be able to scan quickly over a band. Also, when transmissions are on distinct channels, it is often useful to step in increments of single channels. For example short wave broadcast stations are spaced 5 kHz apart, and many other transmissions are also channelized.

Some synthesized sets have memories and these can be used to store particular frequencies which are monitored regularly. Some radios have about 10 to 15 memories whereas others may have more.

4.1.2 AF and RF gain

All receivers have at least one control to alter the gain. The most common is an AF or audio frequency gain control. This is a volume control and is set to give sufficient volume of listening.

The other type of gain control is called an RF gain control. This is used to alter the gain of the radio frequency stages. When strong signals are present, it is sometimes necessary to reduce the gain of the early stages of the set to prevent overloading and the resultant distortion.

Some older sets may even have an IF gain control. These are rarely seen these days and they are used in conjunction with the RF gain control to ensure that none of the following stages become overloaded.

Instead of an RF gain control some sets have an RF attenuator. This performs essentially the same function as an RF gain control but it acts right at the input to the receiver to prevent even the first RF amplifiers from becoming overloaded.

4.1.3 Automatic gain control

The strength of signals picked up by any radio set varies enormously. Some may be over a million times stronger than others. Often the strength of a signal may vary as the signal fades in and out. To help accommodate these changes and to save having to continually vary the gain, a circuit known as an **automatic gain control** or AGC is used in virtually all sets today. This circuit monitors the level of the signal in the later stages of the set and applies a signal to the earlier stages to reduce the amplification. In many sets this circuitry enables the RF gain control to be totally eliminated.

On some receivers there is a choice between slow and fast AGC times. On modes like SSB where the level of the signal changes even during the course of a word, the best results are achieved if the AGC follows the general trend of the signal. With the AGC set to a slow time it may take a second or more for the receiver to reach its full gain level once the signal has been removed. A fast AGC is more appropriate for a Morse signal where the level of the transmitted signal is almost constant.

4.1.4 Beat frequency oscillator

A beat frequency oscillator (BFO) or carrier insertion oscillator (CIO) is used when receiving single sideband and Morse signals. In many sets there is a separate control to switch it on and possibly one to alter its frequency slightly. On other sets there may be a mode switch and the BFO is automatically switched on for Morse and single sideband. Where upper sideband and lower sideband are marked on the switch the frequency of the oscillator is set automatically.

In sets where the frequency of the BFO can be adjusted it needs to be set for optimum performance. The signal being received should be tuned in for the best reception by placing it in the middle of the receiver passband. The setting of the oscillator is then adjusted to suit. It will be slightly to the side of the signal being received as shown.

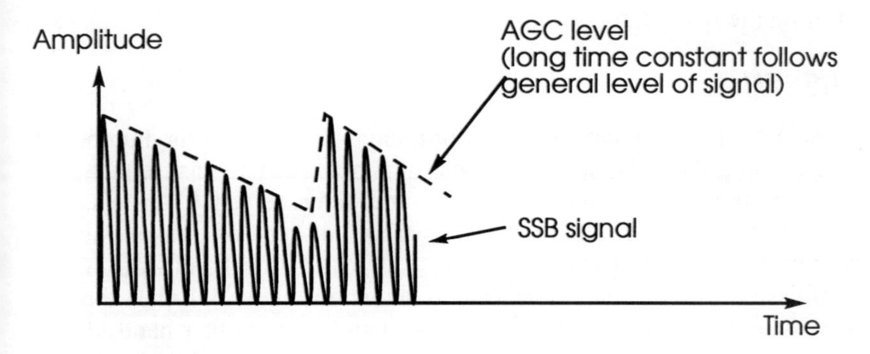

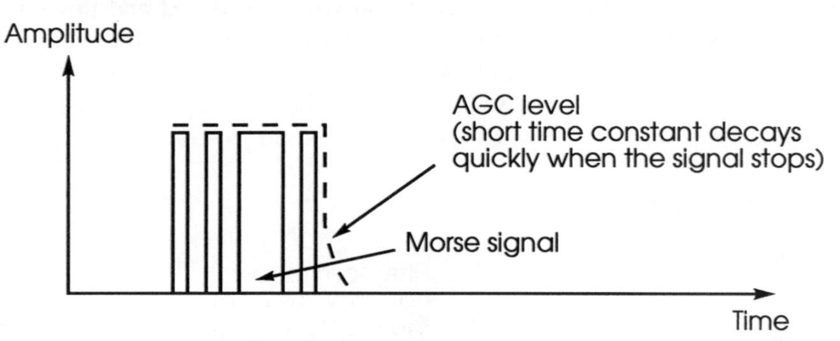

Figure 4.2 *AGC response times*

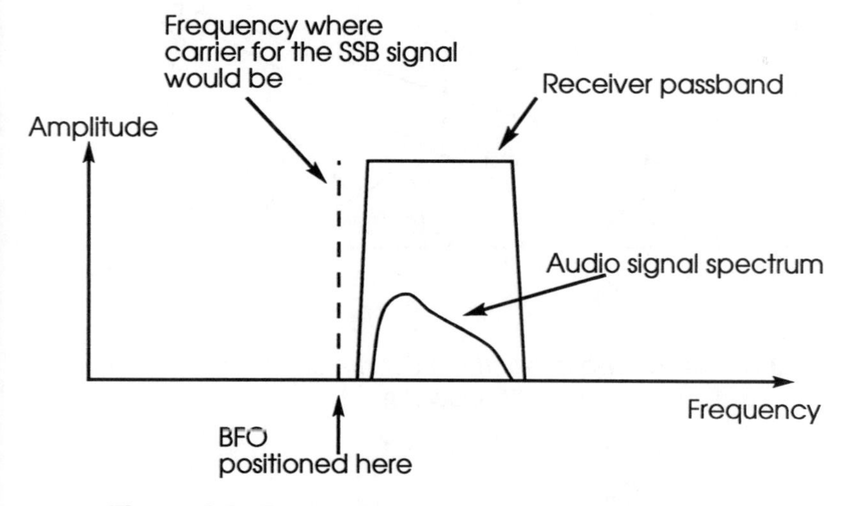

Figure 4.3 *Positioning of a beat frequency oscillator*

Once the BFO has been set up for a particular mode it is just a matter of tuning the main dial until the signal becomes intelligible.

4.1.5 Filters

The selectivity of a receiver is of prime importance. With the number of stations using the radio spectrum today it is necessary to ensure that only signals on the selected frequency or channel are received. Transmissions have a finite bandwidth and this varies from one type of transmission to another. This means it is necessary to match the filter bandwidth to the transmission. If the filter is too wide as shown in Figure 4.4, then there is the possibility of picking up unwanted signals. On the other hand, if it is too narrow then part of the wanted signal will be rejected and this will result in distortion. It is therefore necessary to use filters of the required bandwidth. Typically for a short wave AM transmission a bandwidth of 6 kHz is about standard. For SSB, 2.7 kHz or thereabouts is used, and for Morse, bandwidths of 500 or 250 Hz are often used.

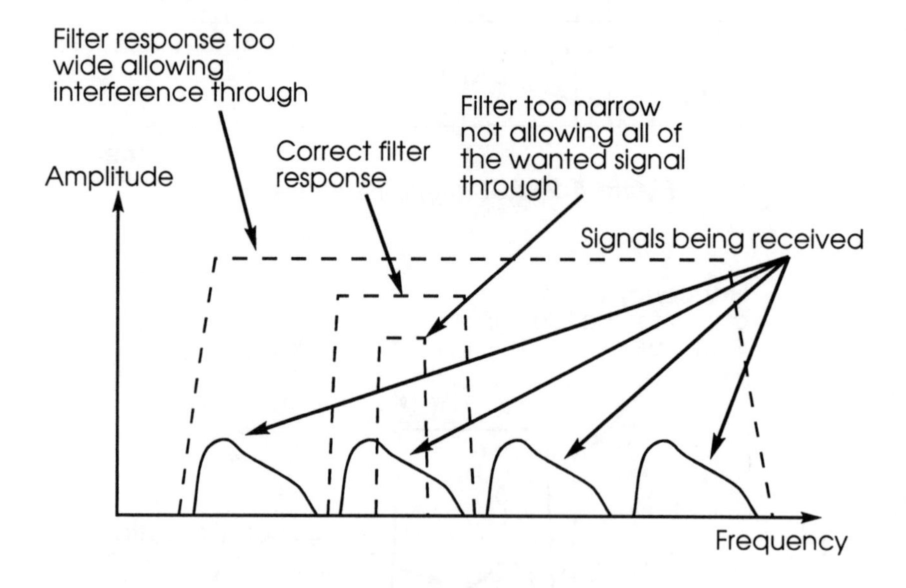

Figure 4.4 *Filter bandwidths must match the transmission bandwidth*

4.1.6 Noise limiter

Electrical noise like that from car ignition systems, electric motors and a host of other appliances consists of short impulses of electrical energy which limit the effective sensitivity of a receiver.

A number of circuits exist to limit the effect of this type of interference. The simplest ones limit the amplitude of the pulses and provide a small degree of improvement. The more sophisticated circuits can actually remove the noise impulses leaving the signal far more readable.

There is usually a control associated with the noise limiter. Its operation will depend on which type of circuit is in use. The simpler limiter types should be adjusted until the volume of the signal just starts to be reduced. Other types may require adjusting so that the noise is cancelled or nulled out. This requires adjustment of the control until the level of the noise reaches a minimum.

4.1.7 Squelch

Most short wave sets do not have a squelch control. They are normally found on scanners intended primarily for use on frequencies above 30 MHz. Their function is to mute the receiver when no signals are present. They are particularly effective when receiving FM signals. When switched to this mode a receiver normally produces a large amount of background hiss or noise when no signal is present. The squelch circuitry mutes the audio circuitry to remove this noise under these conditions.

When a squelch control is present, it should be set so that it just mutes the receiver when no signal is present. By setting the control in this position even the weakest signals will be heard.

4.2 External connections

There are a number of external connections to any receiver. Power, whether it is a mains or low voltage supply, may be needed, unless batteries are used. External speakers and headphones may also be required.

The supply is the first to be considered. Most communications sets use a mains supply. The connections for this are normally self-explanatory, but bear in mind that the correct value fuse should be used. For most sets this will be a low value, and certainly not the 13-amp variety found in most new plugs in the UK. The correct value should be fitted to give the maximum protection.

Some sets may need an external DC low voltage supply. When setting this up, care is needed to install the supply so that there is no possibility of the polarity of the supply ever being reversed. If this happened serious damage may result to the set.

An external speaker may be required. An 8-ohm speaker will operate perfectly satisfactorily in most applications, but it is best to check the exact requirements.

Headphones may be required. The standard types used with most portable radios and cassette players may well suffice, but again it is best to check in the manual.

4.3 The superhet

To make the most of any receiver it is necessary to have a basic idea of the internal operation. In this way it is possible to use the controls to their best advantage. Although there are a number of different types of receiver, the one which is almost universally used is called a **superhet**.

This type of receiver was first developed towards the end of the First World War and gained popularity in the 1920s and 1930s. Called the superhet, it operates by changing the frequency of the incoming signal to a fixed intermediate frequency. Here the signal is filtered and amplified before being demodulated. Then the resulting audio is amplified in the normal way.

Figure 4.5 *Like most receivers today this 'World Band' receiver uses the superhet principle*

4.3.1 Mixing

The whole idea of the superhet revolves around the process of mixing. Unlike ordinary audio mixers which simply add the signals together in a linear fashion, radio frequency mixers multiply the two signals together. When this is done, new signals are generated at frequencies equal to the sum and difference of the two original ones. In other words, if two signals with frequencies equal to f_1 and f_2 are mixed together, two new signals at frequencies equal to $f_1 + f_2$ and $f_1 - f_2$ are produced. This can be seen a little more easily if real figures are used. If the two original signals are at frequencies of 600 kHz and 800 kHz, then the two new signals will be 1400 kHz ($f_1 + f_2$) and 200 kHz ($f_1 - f_2$).

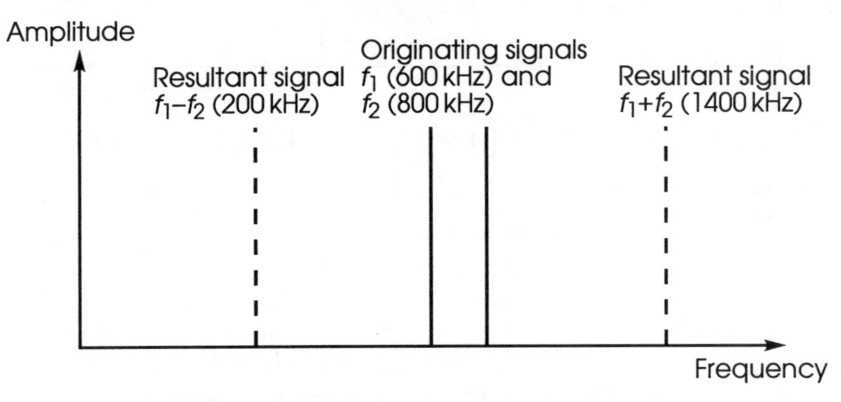

Figure 4.6 *Signals mixed or multiplied*

Using the mixing or multiplying process it is possible to change the frequency of a signal up or down in frequency. To illustrate this take the example of a signal at 2.0 MHz. This mixes with the local oscillator at 1.5 MHz to give a signal at the intermediate frequency of 0.5 MHz or 500 kHz. This signal represents the difference mixer product. The sum appears at 3.5 MHz and is easily filtered out as it is very well removed from the frequency of the filters.

Unfortunately there is another mixer product which can pass through the intermediate frequency stages. The first signal represented the difference between the incoming signal and the local oscillator. However a signal representing the local oscillator minus the incoming signal can also pass through the filters. Taking the same example again where the local oscillator frequency is 1.5 MHz, an incoming signal of 1.0 MHz also gives an output at 500 kHz.

It is obviously undesirable to have two signals on totally different input frequencies able to pass through the intermediate frequency filters and amplifiers. Fortunately it is relatively easy to remove the unwanted or **image** signal to leave only the required one. This is achieved by placing a tuned circuit in the radio frequency stages to remove the signal as shown in Figure 4.7.

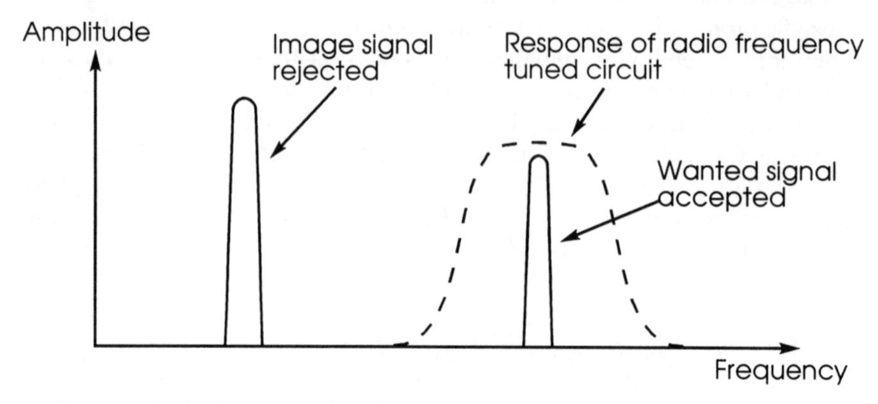

Figure 4.7 *A tuned circuit used to reject the unwanted image signal*

From the diagram it can be seen that this tuned circuit does not need to reject signals on adjacent channels like the main intermediate frequency filters. It is only necessary to reject the image signal. This happens to fall on a frequency equal to twice the intermediate frequency away from the wanted signal. In the example this is 1 MHz away from the wanted signal. This is compared to stations on adjacent channels which are rejected by the main intermediate frequency filters which may be a few kilohertz away, or in some cases less.

Tuning the receiver is accomplished by changing the frequency of the local oscillator. If the oscillator is moved up in frequency by 100 kHz then the frequency of the signals which are received will be 100 kHz higher. In the example used earlier if the local oscillator is 100 kHz higher at 1.6 MHz then signals will be received at 2.1 MHz, assuming the higher frequency signals are wanted and those at 1.1 MHz are rejected as the image.

If the local oscillator and hence the receiver frequency is tuned then it is necessary to ensure that the RF tuning also moves at the same rate. In this way the RF tuning will be tuned to the correct frequency and will ensure the wanted signal passes through without being attenuated. It also ensures that the image signal is rejected.

To ensure that this occurs, the tuning for the local oscillator and RF circuits must be linked so that they track together. Where mechanical tuning capacitors are used, two sections of the capacitor are physically linked so that they vary at the same rate. In this way the two tuned circuits can be varied by the same degree at the same time. In more modern receivers electronic methods of tuning are used and these are designed to track at the same rate to maintain the correct circuit conditions.

4.3.2 Overall operation

A block diagram of basic superhet radio is shown in Figure 4.8. The signals enter the radio frequency circuits where the required band of frequencies is selected and those which might give rise to image signals are rejected. In the same stages the signals are amplified to the required level and passed into the mixer or multiplier circuit where they are mixed with the signal from the local oscillator. This converts the incoming signals to the intermediate frequency. As the mixer performs the action of changing the frequency of the received signal, it is sometimes called the **frequency changer**.

Once at the intermediate frequency, the signals are amplified again. They are also filtered using fixed frequency filters to remove unwanted signals which are off channel. After they have passed through the intermediate frequency stages, they need to be converted from a radio frequency signal to an audio frequency one. This is accomplished using a demodulator. To accommodate the different types of transmission, different types of demodulator may be needed and these can be switched in as necessary. For amplitude modulation a simple diode detector like that shown in Chapter 2 is all that is required. SSB and Morse require a beat frequency oscillator and product detector. The BFO reinserts the carrier in the case of SSB. For a Morse signal it has exactly the same action but is best visualized as beating with the incoming signal to produce the required audio tone. The product detector is a mixer where the beat frequency oscillator signal and the received signal are mixed to give the required audio output.

Once the audio signal has been generated, this is passed into the audio amplifier where it is amplified and passed into a loudspeaker or headphones in the usual way.

4.3.3 Image response

Although the image response has already been mentioned, it is a very important feature of any superhet radio and needs further explanation. An image signal can enter the IF stages, be amplified and filtered with the wanted signals despite the fact that there is RF tuning. On many receivers the problem of image signals becomes worse at the top end of their coverage. In any receiver the difference between the image frequency and the wanted frequency is twice the intermediate frequency. At the bottom end of the coverage this gives a large percentage

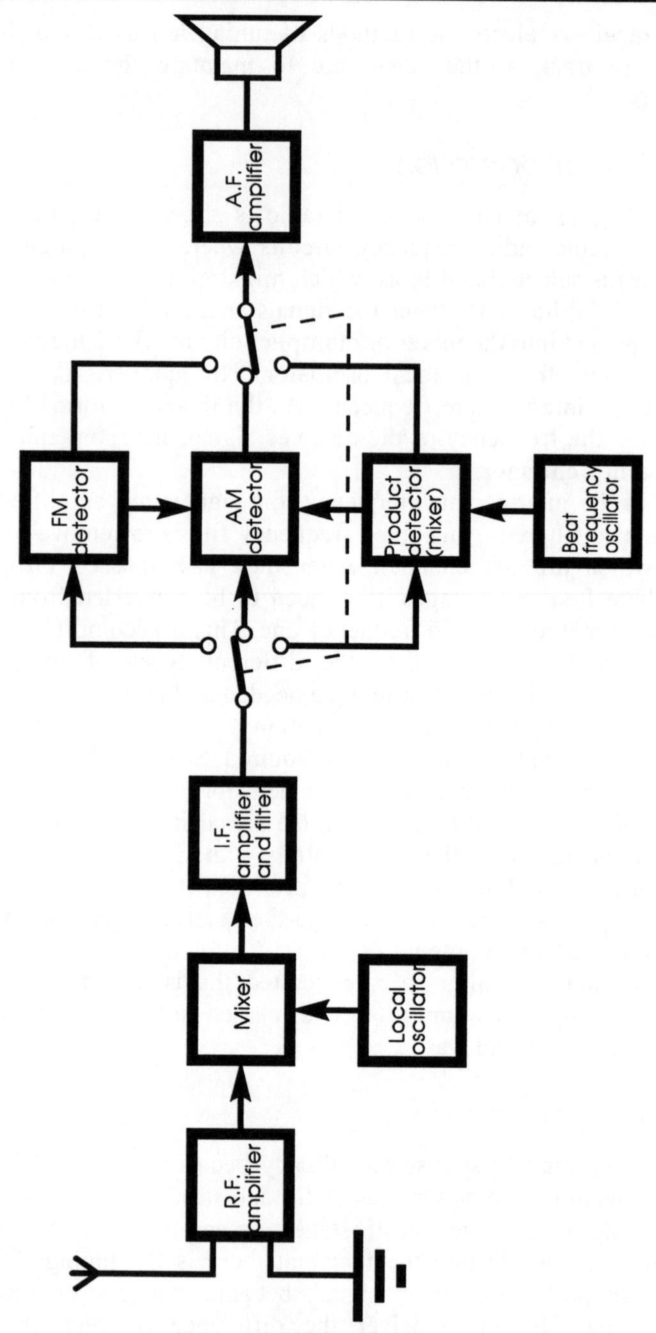

Figure 4.8 *Block diagram of a basic superhet*

difference between the two signals whereas at the top end this is smaller and the degree of rejection of the image signal will be less.

To see the effect of this, take an example of a receiver with an intermediate frequency of 455 kHz. If the local oscillator is tuned to the low side of the signal as shown in Figure 4.9 and the receiver is tuned to 2 MHz then the image frequency is at 1.09 MHz. As seen in the diagram, a high degree of rejection is achieved. However, when the set is tuned to 30 MHz, the image appears at 29.09 MHz. The bandwidth of the filter will remain about the same in terms of the percentage of the frequency of operation. However, it is much greater in absolute terms. As a result the image signal is rejected less at higher frequencies than lower ones.

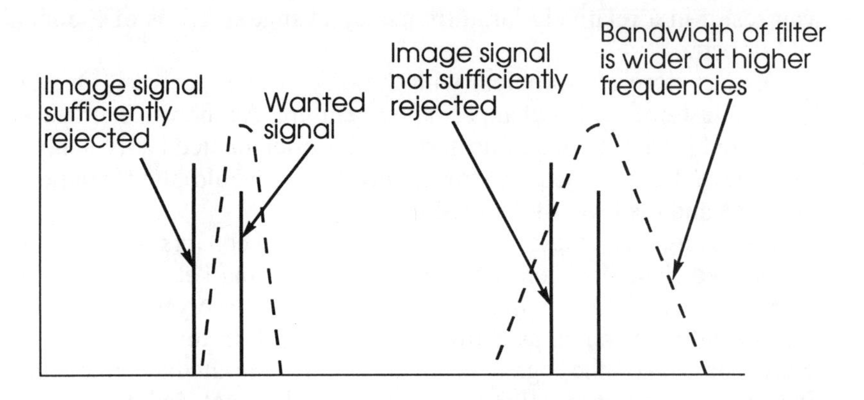

Figure 4.9 *Image rejection when the receiver is tuned to different frequencies*

Obviously one method of improving the image response is to make the front end RF tuning sharper. This is not ideal because it means that the local oscillator and RF tuning have to track more accurately. Any discrepancy is then likely to lead to a reduction in the level of the wanted signal.

A more suitable way of overcoming the problem is to increase the intermediate frequency. This has the effect of increasing the difference between the wanted and image signals, making filtering in the RF stages easier. In fact in some modern receivers the incoming signal is actually converted to a higher frequency first.

However, before looking at these sets it is worth noting how the image response is specified. It is specified as a given number of decibels at a particular frequency. For example it may be 50 dB at 30 MHz. This indicates that if signals of the same strength on the image and wanted frequencies were present at the input, then at the output of the receiver the image one would be 50 dB lower than the wanted one.

4.3.4 Multiple conversion sets

There are a number of problems with the superhet principle in its basic form. Firstly, its image rejection may be poor, especially at high frequencies as previously described. If the intermediate frequency is increased to improve this, then the selectivity may suffer.

A further problem can be the stability of the set. To tune the receiver, the local oscillator has to be made variable. In turn this means that the receiver is prone to drift once it has been set to a particular frequency. This can be particularly annoying when the set has to be regularly adjusted to keep it tuned to a station. Like image rejection, drift is more of a problem at higher frequencies. Here a very small percentage change can result in a relatively large frequency change in terms of the number of kilohertz.

Drift is more of a problem with the older valve receivers. One of the major causes of drift is changes in temperature. As the valves themselves generated large amounts of heat, these sets often drifted badly. Only after they have been turned on for a few hours would their temperature stabilize and the level of drift reduce.

To overcome these problems multiple conversion sets were introduced. The first ones to appear on the market used a variable frequency oscillator for the first conversion and a crystal oscillator for the second conversion, as shown in Figure 4.10. In this scheme the first conversion takes the signal down to a high intermediate frequency to increase the difference between the wanted signal and the image. A second conversion using a fixed frequency oscillator is used to convert the signal down to the second intermediate frequency stage. This oscillator is on a fixed frequency and is usually based on a quartz crystal for convenience and stability.

As the first intermediate frequency was higher than the second, a major improvement in image response was seen. Unfortunately the problem of drift still remains as the variable frequency operates at a high frequency.

One method of overcoming the drift problem is to use a crystal controlled first conversion as shown in Figure 4.11. As the first oscillator is crystal controlled, it is much more stable. The first conversion is used to convert a band of frequencies down to the first IF which may have a bandwidth of about half a megahertz.

A variable frequency oscillator is used to convert the signals down to the fixed frequency IF stages where the filtering is performed. As this oscillator is not switched and is running at a much lower frequency it can be made very stable. The higher frequency crystal oscillator is inherently more stable and gives the overall set much better stability.

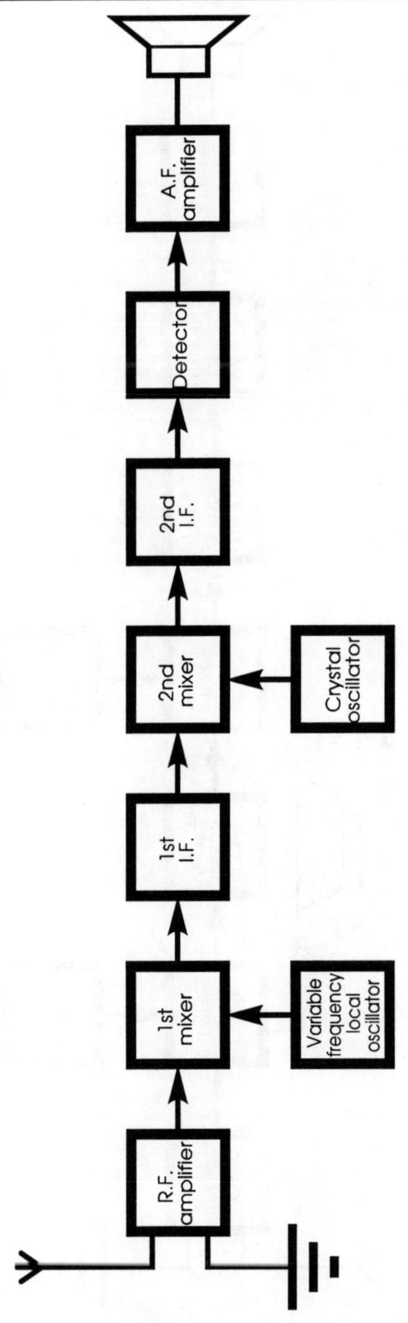

Figure 4.10 *A double conversion superhet with variable frequency first oscillator*

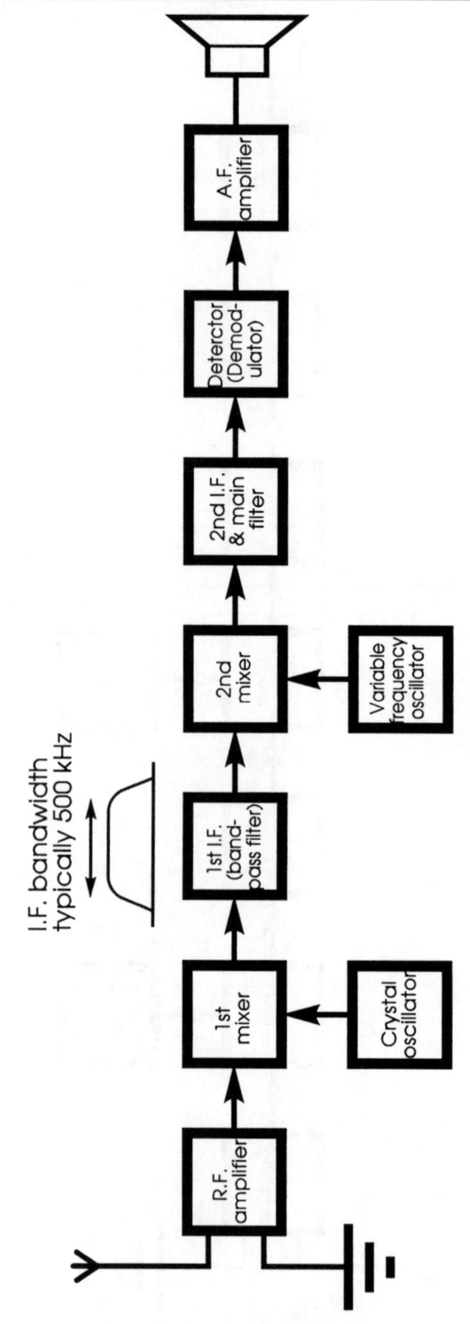

Figure 4.11 *A double conversion superhet with crystal controlled first conversion*

The major drawback of this type of design is that it needs a large number of crystals in the first conversion if it is to cover a wide spread of frequencies. However the solution was ideal for amateur band only receivers where only a limited number were needed.

4.3.5 Synthesizers

To overcome the problems of stability, most receivers today use frequency synthesizers for their local oscillators. Many receivers using synthesizers will boast such terms as 'PLL', 'Quartz' and 'Synthesized' in their specifications or advertising literature.

Frequency synthesizers offer very high degrees of stability. They can operate over a wide range of frequencies and they interface easily to digital circuitry. This enables them to be controlled by the microprocessors often found in today's sets so that they can provide new degrees of flexibility and many more facilities. Push button tuning together with the normal tuning knob, frequency memories, scanning and a whole host of other facilities are now available.

A frequency synthesizer uses a **phase locked loop** (PLL) as the basis of its operation. Phase and comparing the phase of two signals is crucial to the operation of this circuit.

The **phase** of a signal is the position within a cycle. Often this is likened to the point on the signal travelling around a circle with a complete cycle being equal to 360 degrees. Each point on the sine wave has an equivalent point on the circle. If there are two signals, they may not be at the same point in the cycle, as shown in Figure 4.12. There is a **phase difference** between them. Each waveform will be at a different point on the circle and the phase difference is equal to the angle between the two points on the circle.

A basic phase locked loop is shown in Figure 4.13. There are three basic circuit blocks: a phase comparator, a voltage controlled oscillator and a loop filter. A reference oscillator is sometimes included in the PLL. This is not strictly part of the basic loop even though a reference signal is required for its operation.

The loop operates by comparing the phase of two signals. The signals from the voltage controlled oscillator and reference enter the phase comparator which produces a third signal equal to the phase difference between the two. This is passed through the loop filter. This performs a number of functions within the loop, but in essence it removes any unwanted products from the error voltage before the tune voltage is applied to the control terminal of the voltage controlled oscillator. The error voltage tries to reduce the discrepancy between the two signals entering the phase comparator. Consequently the voltage controlled

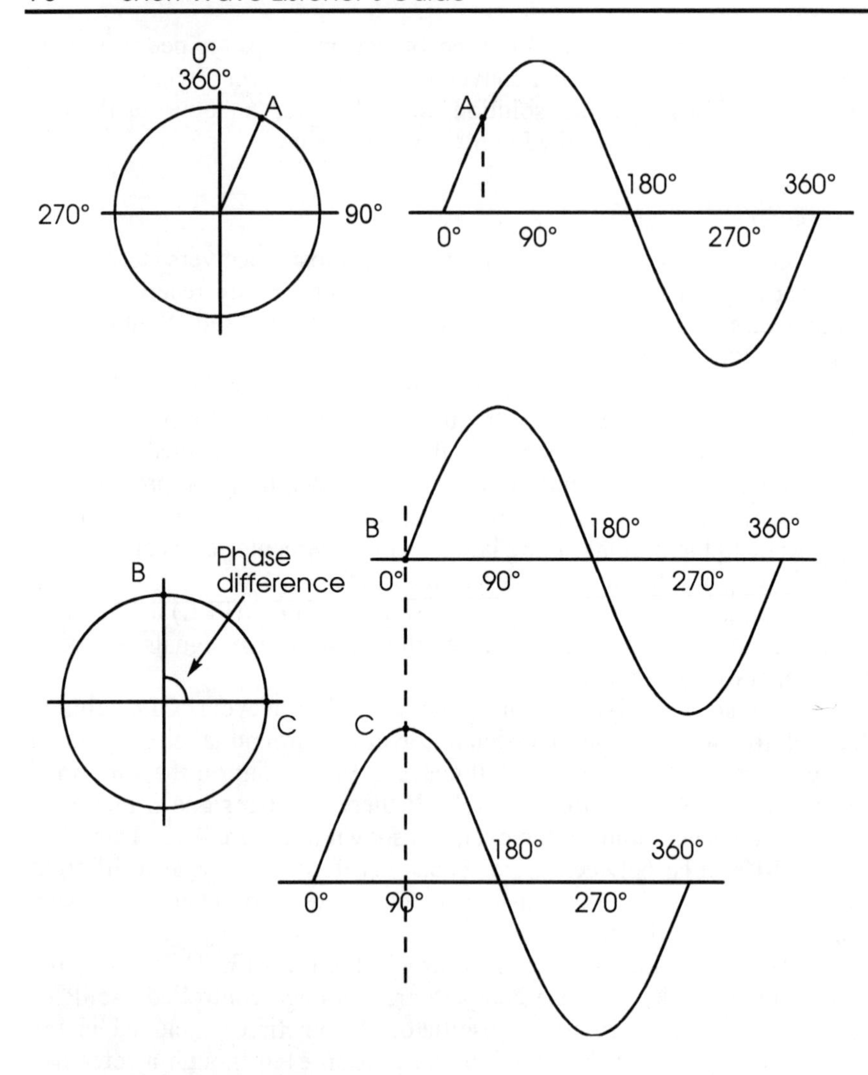

Figure 4.12 *Phase difference between two signals*

oscillator will be pulled towards the frequency of the reference. When 'in lock' there will be a steady state error voltage. This is proportional to the phase error between the two signals. Only when the phase difference between the two signals is changing is there a frequency difference. As there is no phase change when the loop is in lock, the frequency of the voltage controlled oscillator is the same as the reference.

A phase locked loop needs some additional circuitry if it is to be converted into a frequency synthesizer. A frequency divider is added into the loop between the voltage controlled oscillator and the phase

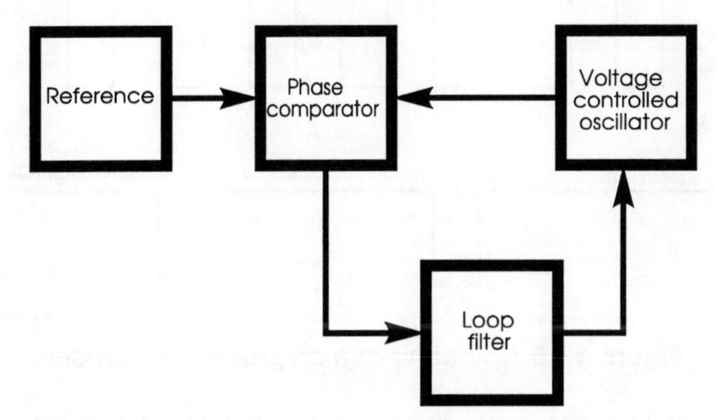

Figure 4.13 *Block diagram of a basic phase locked loop*

comparator as shown in Figure 4.14. The divider divides the frequency of the incoming signal by a certain ratio. If it is set to two, then the output frequency is half the input and so forth.

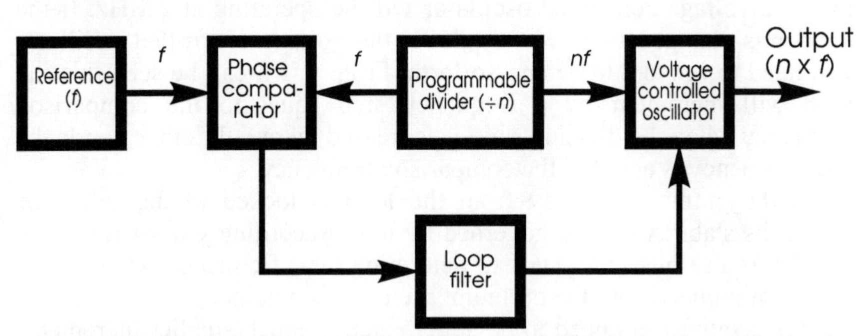

Figure 4.14 *A programmable divider added into a phase locked loop enables the frequency to be changed*

Programmable dividers are used in a variety of applications, including many radio frequency uses. Essentially they take in a pulse train like that shown in Figure 4.15 and give out a slower train. In a divide by two circuit only one pulse is given out for every two that are fed in and so forth. Some are fixed, having only one division ratio. Others are programmable. Digital or logic information can be fed into them to set the division ratio.

When the divider is added into the circuit the loop still tries to reduce the phase difference between the two signals entering the phase comparator. When the circuit is in lock, both signals entering the phase

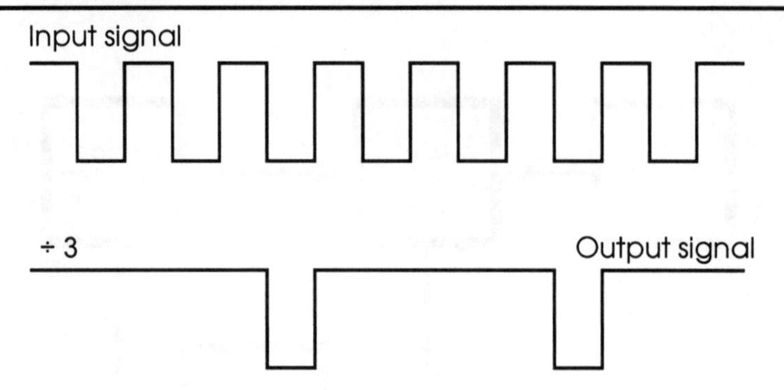

Figure 4.15 *Operation of a programmable divider*

comparator have the same frequency. For this to be true the voltage controlled oscillator must be running at a frequency equal to the phase comparison frequency times the division ratio as shown in the diagram.

Take the example of the divider set to two and a reference frequency of 1 MHz. When the loop is in lock, the frequency entering both ports of the phase comparator will be 1 MHz. If the frequency divider is set to two, the voltage controlled oscillator will be operating at 2 MHz. If the divider is changed to divide by three, the voltage controlled oscillator will need to run at 3 MHz and so forth. From this it can be seen that the loop will increment by a frequency step equal to the comparison frequency when the division ratio is increased by one. In other words the step frequency is equal to the comparison frequency.

As the output frequency from the loop is locked to the reference signal, its stability is also governed by this. Accordingly most reference oscillators use quartz crystals to determine their frequency. In this way drift is minimized, and the optimum accuracy is obtained.

Most synthesizers need to be able to step in much smaller increments if they are to be of any use. This means that the comparison frequency must be reduced. This is usually accomplished by running the reference oscillator at a frequency of 1 MHz or so, and then dividing the signal down to the required frequency using a fixed divider. In this way a low comparison frequency can be achieved. The block diagram of a very basic synthesizer is shown in Figure 4.16. Those used in many receivers are more complicated, using more than one synthesizer loop. By combining them all together a high performance oscillator can be made.

4.3.6 Direct digital synthesizers

Although synthesizers based upon phase locked loops are the most common, they are not the only type. A form of synthesizer called a

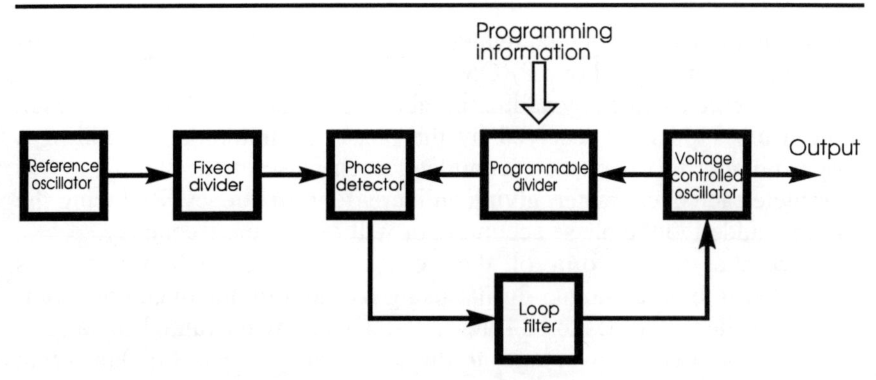

Figure 4.16 *Comparison frequency reduced by adding a fixed divider after the reference oscillator*

direct digital synthesizer or DDS is starting to appear in a number of the top end receivers and amateur radio transceivers. In most of today's designs they are used in conjunction with a phase locked loop to give high performance at a reasonable cost.

This type of synthesizer stores a digital representation of the signal in a memory which is regenerated to form a sine wave. The basic synthesizer consists of four basic blocks as shown in Figure 4.17. These are a phase accumulator, waveform map, a digital to analogue converter and finally a filter to remove out of band signals.

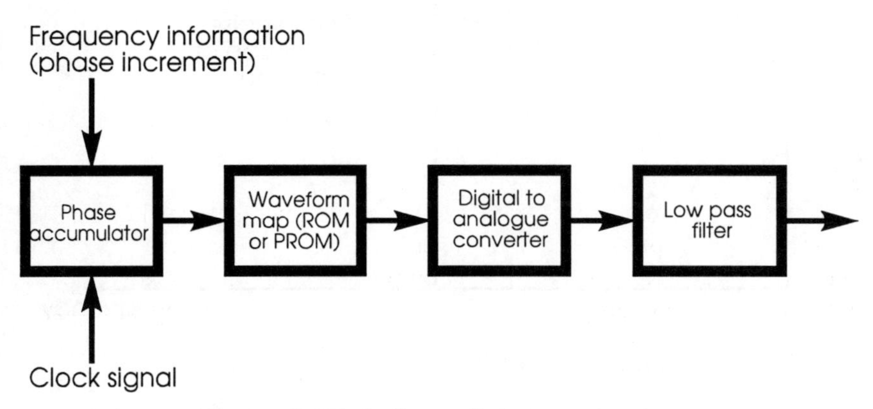

Figure 4.17 *A direct digital synthesizer*

The phase accumulator is basically a form of counter. Numbers can be added into the counter, corresponding to an increase in the phase or position in the cycle. Once it reaches its maximum value, the counter resets and returns to zero.

The information from the phase accumulator is used to determine the actual voltage in digital terms from a waveform map. For each value of phase there must be an equivalent digital value for the output. This circuit is either a ROM or a PROM.

The circuit operates by adding numbers to the phase accumulator each time a clock pulse is received by the phase accumulator. By adding a larger number to the phase accumulator each pulse the accumulator will complete each cycle faster, giving an increase in frequency. Reducing the number added to the phase accumulator will reduce the frequency.

Once the digital form of the voltage has been determined it is converted into an analogue format using a digital to analogue converter. After this the signal has to be filtered to remove any unwanted signals.

There are several advantages to direct digital synthesizers. One is that they can switch very quickly from one frequency to another. Another is that they have relatively low levels of phase noise.

4.3.7 Phase noise and reciprocal mixing

One of the main problems with frequency synthesizers is that they can generate high levels of phase noise if care is not taken in the design. This noise is caused by small amounts of phase jitter on the signal and it manifests itself as noise spreading out either side of the signal as shown in Figure 4.18.

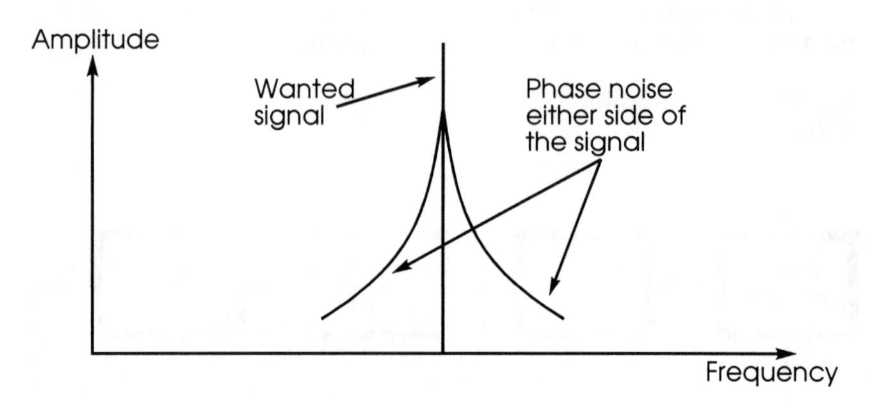

Figure 4.18 *Phase noise on a signal*

Any signal source will have some phase noise. Crystal oscillators are very good and free running variable frequency oscillators normally perform well. Unfortunately synthesizers do not always fare so well and this can adversely affect the radio performance in terms of the reciprocal mixing performance.

This can be explained by considering a receiver tuned to a strong signal. The local oscillator output mixes with the wanted station to produce a signal which falls within the passband of the receiver filters. If the receiver is tuned off channel by a given amount, for example 10 kHz, then the station can mix with the local oscillator phase noise 10 kHz away from the main oscillator signal as shown in Figure 4.19. If the incoming station signal is very strong, the reciprocal mixing effect can mask out weaker stations, thereby reducing the effective sensitivity of the set in the presence of strong off channel signals.

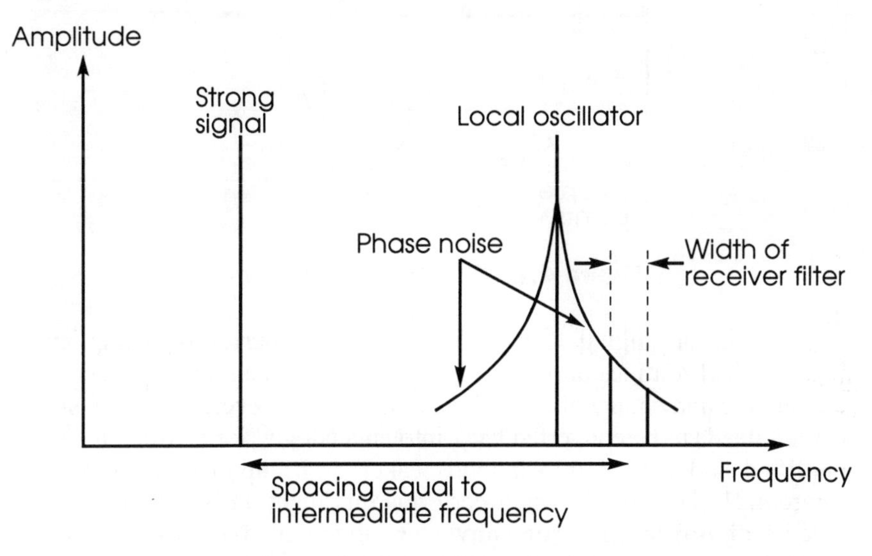

Figure 4.19 *Reciprocal mixing*

4.4 Digital signal processing

Microprocessors are finding uses in an increasing number of applications. It is therefore hardly surprising to find that they are being used in applications which are normally performed by analogue electronics. Functions such as filtering and demodulation can all be performed digitally by specialized digital signal processors. As a result some of the higher priced receivers are beginning to use these techniques and it is expected that in the coming years this process will become more widespread.

The process builds up a representation of the signal in a digital form by sampling the voltage level at regular time intervals and converting the voltage level at that instant into a digital number proportional to the voltage as shown in Figure 4.20. This process is performed by a circuit called an **analogue to digital converter**, often called an A to D converter or ADC.

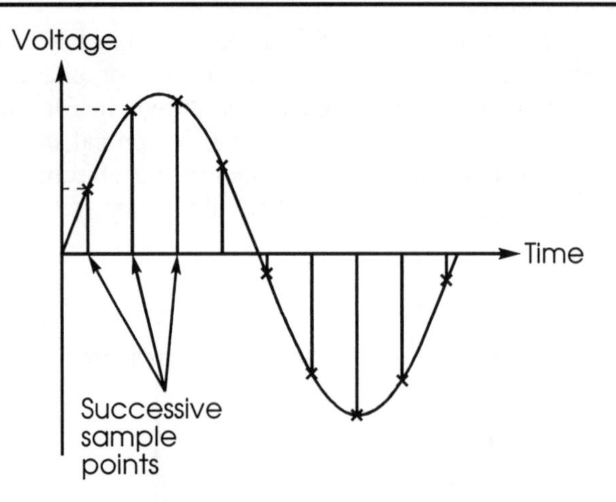

Figure 4.20 *Sampling a waveform*

Once in a digital form the processor performs complicated mathematical routines upon the representation of the signal. These filter, demodulate and change the signal as required. However, to use the signal again it needs to be converted back into an analogue form where it can be amplified and passed into a loudspeaker or headphones as shown in Figure 4.21. The circuit which performs this function is not surprisingly called a **digital to analogue converter**, often called a D to A converter or DAC.

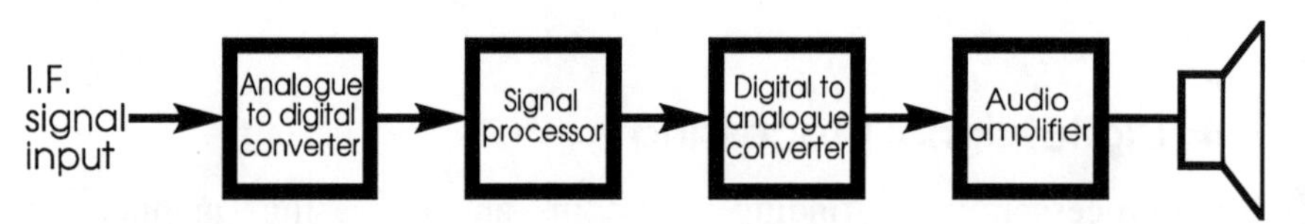

Figure 4.21 *Block diagram of a digital signal processor*

The advantage of digital signal processing is that once the signals are converted into a digital format they can be manipulated mathematically. The signals can be processed more accurately, and this enables better filtering, demodulation and general manipulation of the signal. Unfortunately it does not mean that filters can be made with infinitely steep sides because there are mathematical limitations to what can be accomplished.

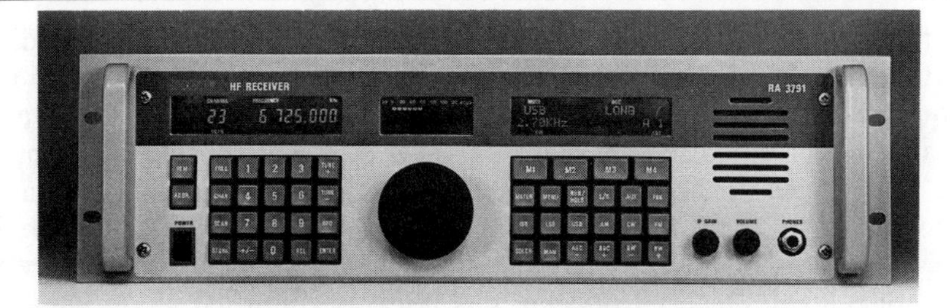

Figure 4.22 *A professional receiver which uses digital signal processing (photo courtesy Racal Communications Systems Ltd)*

4.5 Synchronous AM detection

Improvements in amplitude modulation reception can be achieved by using a system known as **synchronous detection**. Essentially it uses a beat frequency oscillator and product detector or mixer for AM. For this to work the beat frequency oscillator must be kept on the same frequency or zero beat with the carrier. There are a number of methods of doing this, but the one which is most commonly used employs a high gain limiting amplifier. This takes the AM signal and removes the modulation to leave only the carrier as shown in Figure 4.23. This is used as the beat frequency oscillator to demodulate the signal.

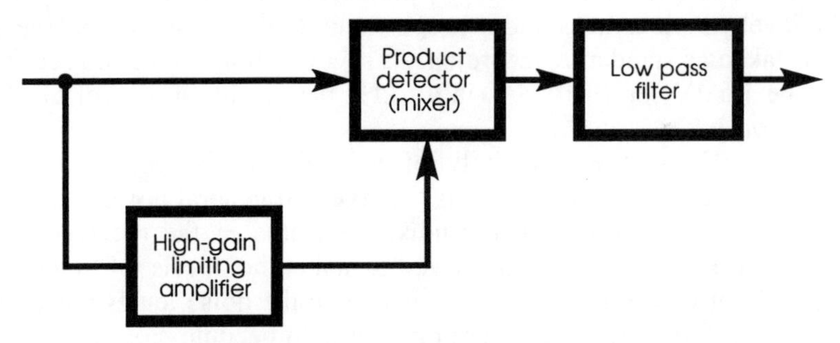

Figure 4.23 *Synchronous detector*

Synchronous detection of AM gives a more linear method of demodulation. It also gives more immunity against the effects of selective fading on the short wave bands.

4.6 Sensitivity

One of the main requirements of any receiver is that it must be capable of amplifying the signals which are picked up by the aerial so that they are sufficiently strong to be demodulated. There is no problem in adding sufficient amplifying stages to enable the signal to be brought up to the required level. The problem is that within every stage of the receiver a small amount of noise is introduced. This does not cause a problem in the later stages of the set. Here the signal levels are high and the small amounts of noise are well below the signal levels. The problem occurs in the first stages of the set. Here the signal levels are low, comparable to the noise level. Once introduced, this noise is amplified along with the wanted signal. Consequently the first stage in the receiver is the most critical. It is imperative that noise levels here are kept to an acceptable level.

It is the levels of noise which limit the sensitivity of a set. As a result, the sensitivity is specified in terms of its noise performance. There are several methods used to measure the sensitivity of a set. For most short wave receivers for frequencies below 30 MHz the ratio of a signal of a given level to the noise is used. A ratio of 10 dB is normally taken as the signal to noise ratio as this gives a 10:1 ratio in levels. As the level of noise is proportional to the bandwidth, this also has to be included in the specification. Typically a specification of 0.5 μV for 10 dB S/N signal to noise in a 3 kHz bandwidth might be expected. This would be for the performance of the receiver for SSB where the beat frequency oscillator is switched in.

If the level is to be specified for an amplitude modulated signal, the level of modulation also has to be included. This is because the final audio level which is measured is proportional to this. Normally a level of 30% is taken as standard. Accordingly a specification for a good receiver might be 1.5 μV for 10 dB S/N in a 6 kHz bandwidth at a level of 30% modulation.

Above 30 MHz the sensitivity of the receiver becomes far more important. On the short wave bands the levels of general noise picked up by the aerial will mask out any noise generated in the receiver. This means that it is pointless to strive to make the levels of internally generated noise very low. As frequencies rise the noise levels picked up by the aerial fall and the receiver noise starts to become more important. In view of this a more accurate and flexible system is used. Called the **noise figure** it is a measure of the noise which a complete receiver, unit or system introduces. Typically a short wave receiver may have a noise figure of around 10 dB whereas a VHF or UHF set may have a figure of 3 dB or possibly less.

4.7 Specifying selectivity

With the number of stations on the short wave bands giving rise to high levels of interference, the performance of the IF filter is crucial to the operation of the whole set. Unfortunately it is not possible to make a filter which allows through only the wanted signals and totally rejects all the others. The response of a typical filter is shown in Figure 4.24. From this it can be seen that there are a number of differences when compared to the ideal response.

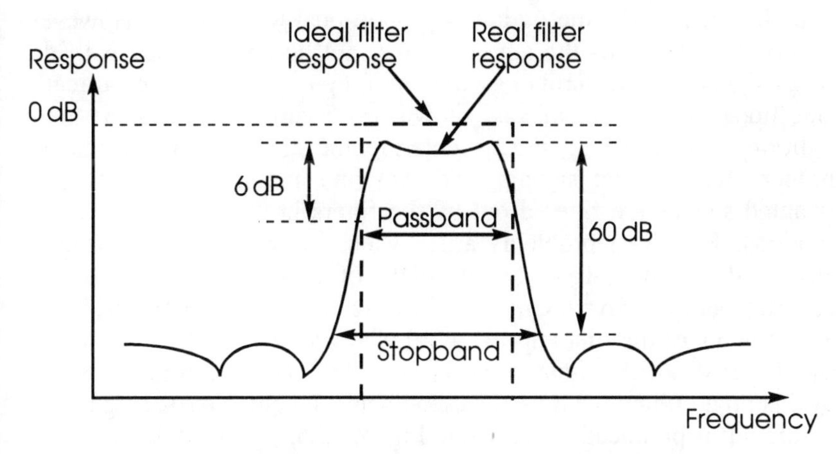

Figure 4.24 *Ideal and real filter responses*

One of the most important features is the filter bandwidth. This is generally measured at the point where the response of the filter has fallen by 6 dB from its mid-band value. However, as other points such as the –3 dB point may be used, this should be specified in its specification. For example, the bandwidth may be given as 2.7 kHz at –6 dB.

The stopband is also of interest. For this the bandwidth for a given attenuation is taken. The normal value for this is –60 dB although other values may be taken. As a result the bandwidth and attenuation levels should be specified. Typically a filter might be specified as having a stopband bandwidth of 6 kHz at –60 dB.

The rate at which the filter reaches its final attenuation is of interest. Even though it may have the correct passband, if it does not reach its final attenuation fast enough, it will let interference through. The shape factor is a measure of the rate at which it reaches its stopband. This is taken as the stopband bandwidth divided by the passband bandwidth. Again the levels of attenuation must be given. Taking the figures already used as an example, the shape factor would be 2.2:1 at 6/60 dB.

4.8 Overload capability

Although the sensitivity of a receiver is very important, it is not the only parameter used in determining the amount of gain required in the set.

Equally important is its ability to withstand strong signals without being overloaded. Surprisingly often it is necessary to listen to a weak station in the presence of many nearby ones which are very much stronger. It is conditions like these which tell the good receivers from the bad.

The problems start to occur as the front end amplifiers cannot tolerate the very strong signals. Under normal conditions the amplifiers should be linear. In other words the output is proportional to the input. However, for very strong signals the amplifier may not be able to cope with the strong signal and it limits, giving an output which is not directly proportional to the input, as shown in Figure 4.25. Under these conditions harmonics of signals entering the amplifier are produced. Another effect is that signals mix with one another. Many of these unwanted signals are filtered out by the filters in the receiver and pass unnoticed. However, problems arise when a harmonic of one signal mixes with another. Take the example of two signals close to the frequency being received. One is at 20 MHz and the other at 20.01 MHz. The harmonic of the first appears at 40 MHz and when it mixes with the second signal, a new signal appears at 19.99 MHz. It is quite possible that this might obliterate a weak station being sought. Not one signal but a whole raft is produced, as shown in Figure 4.26.

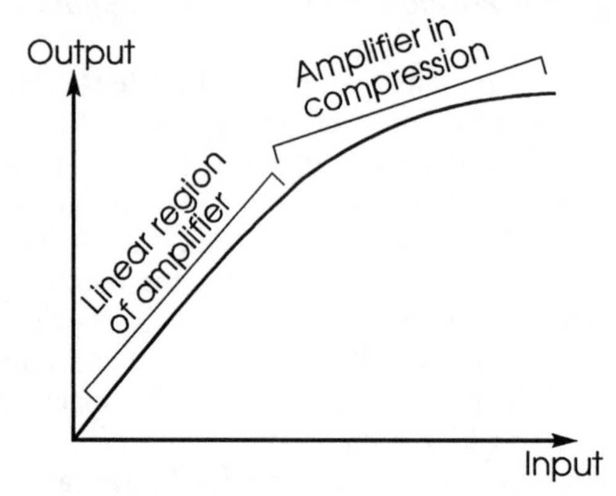

Figure 4.25 *Response of the amplifier circuits*

Other problems can be caused by non-linearity of this nature. One is called **cross-modulation**, and it can be particularly important when listening to broadcast stations. Here the modulation from a strong signal may be heard on weaker signals either side. Naturally this can be very annoying.

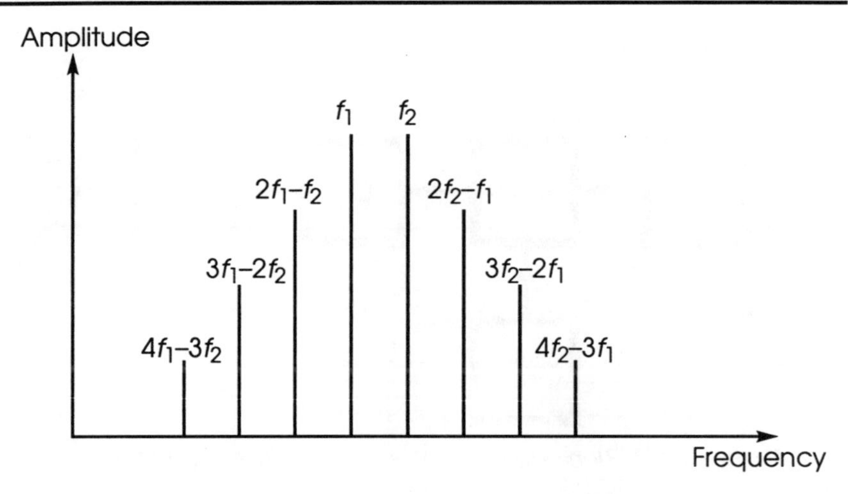

Figure 4.26 *Intermodulation products*

4.9 Direct conversion receivers

Although the superhet receiver is the most widely used type of set, it is by no means the only one. Another type of receiver called a **direct conversion receiver** has a number of advantages and has become popular, especially with home constructors. One of its main advantages is its simplicity when compared to its high degree of performance.

As its name implies, this type of receiver converts its signals directly from radio frequencies down to audio frequencies. Some RF amplification may be provided, but usually most of the gain is supplied by the audio amplifier stages.

The block diagram of a direct conversion of a direct conversion receiver is shown in Figure 4.27. In this the signals first pass into a radio frequency amplifier and tuning stage. This fulfils a number of functions. First of all it naturally amplifies the signals prior to passing them into the mixer. The choice of gain is important, because too much gain can cause the mixer to overload when very strong signals are present. The tuning is also important. If it was not present, then signals from a very wide range of frequencies would appear at the input to the mixer and this again might cause the mixer to overload. This is particularly important where very strong signals may appear at frequencies far away from the ones being used. Finally, the presence of the radio frequency amplifier helps prevent the signal from the local oscillator from reaching the aerial. The amplifier acts as another stage of isolation reducing its level to acceptable limit.

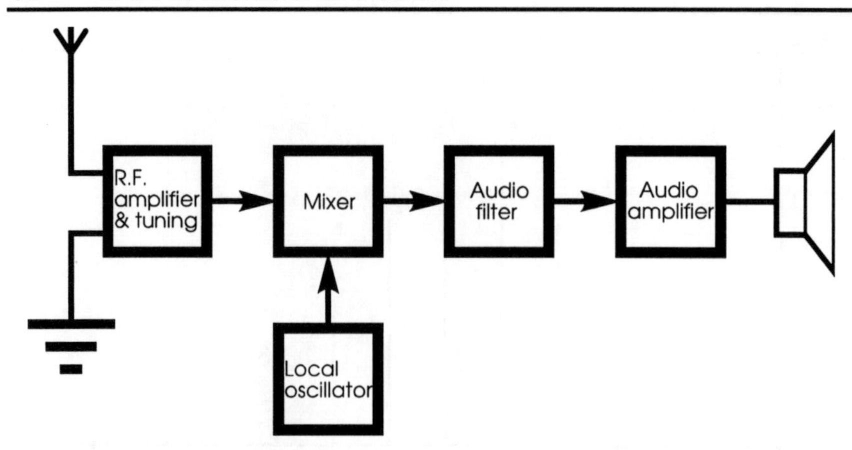

Figure 4.27 *Block diagram of a direct conversion receiver*

The local oscillator circuit is crucial to the performance of the radio. It must be capable of tuning over the range which the radio is required to cover. In some instances this may only be a small band as in the case of a set designed to cover a single amateur band. At other times the set will need to cover a much wider band of frequencies. In either case the design and construction of the oscillator is crucial. One of the prime requirements is that it should not drift. If it does, then the set will need to be retuned at intervals if the same station is required.

Once through the mixer the signals enter the audio amplifier. This section may require some filtering. Off channel signals can produce high frequency interference. This can be reduced by employing an audio filter. For transmissions like Morse a narrow bandpass type of filter can be employed. As the name suggests, this will just allow through a narrow band of signals and is particularly useful for removing other signals which may appear in the audio band. For single sideband reception a low pass filter is provided to remove the unwanted high frequency out of band signals.

One disadvantage of this type of receiver is that it cannot resolve FM signals. It is also not ideal for AM, because the set has to be tuned to zero beat with the signal being received. It is for these reasons that this type of receiver is normally only used by short wave listeners interested in Morse and SSB reception. Another problem is referred to as an **audio image**. This can be demonstrated by tuning through a station with a constant carrier. As the receiver tunes towards the signal a high frequency heterodyne is heard. This steadily reduces in frequency as the receiver tunes closer to the signal. Eventually the set will be zero beat with the signal and then as the set is tuned further away the heterodyne will increase in frequency. From this it can be seen that there are two points where a signal will give a satisfactory beat note, although only one will resolve SSB correctly. However this does mean that interference levels will be higher than for a similar superhet.

Despite these disadvantages, direct conversion receivers give a remarkably high performance for relatively little circuitry. It is for this reason that many radio amateurs have taken to using them when they want to build their own equipment.

5 Aerials

For any short wave listener the aerial or antenna is just as important as the receiver. A poor aerial will severely limit the performance of any receiver, even if it is a particularly good one. It will prevent it from being able to receive the weak and more distant stations. Conversely a good aerial will enable a set to perform to its full potential and enable far more distant stations to be heard.

To be able to choose and install a good aerial it is necessary to have a basic knowledge of some of the principles. Fortunately it is not necessary to have an in-depth knowledge, as the subject becomes very involved. However, from a practical point of view the topic of aerials is very fascinating and can become an interesting addition to the basic hobby of listening.

The function of the aerial is to pick up the electromagnetic radio waves and convert them into electrical signals. Once they exist as electrical signals they are transferred from the aerial element itself into the receiver where they are amplified, filtered and demodulated to give the required audio output.

Aerials have a variety of properties which are of interest to the listener. They can only operate efficiently over a given bandwidth. They also have an electrical impedance and they are polarized, only picking up waves of a certain polarization. Many of these factors are very important and if the aerial is to give the optimum performance it is necessary to ensure it has the correct properties and that it is set up correctly.

5.1 Resonance and bandwidth

An aerial is a form of tuned circuit which appears like an inductor at some frequencies and like a capacitor at others. Like a simple tuned circuit it has a resonant frequency and most aerials are operated at resonance. In view of this, there is only a limited band over which the aerial operates efficiently.

The actual frequency at which the aerial resonates is governed chiefly by its length. The length of the aerial or its elements are generally approximate multiples of a quarter wavelength as the designs of specific aerials will show. This means that the lower the frequency of operation the larger the aerial is likely to be.

The bandwidth is particularly important where transmitters are concerned. If the transmitter is operated outside the bandwidth of the aerial, it is possible that damage may occur. In addition to this, the signal radiated by the aerial may be less for a number of reasons.

For receiving purposes, the performance of the aerial is less critical in some aspects. The aerial can be operated outside its normal bandwidth without any fear of damage to the set. Even a random length of wire will pick up signals and it may be possible to receive several distant stations. However, for the best reception it is necessary to ensure that the performance of the aerial is optimum. Often good aerial systems will enable a receiver to pick up stations at good strength which are totally inaudible on a receiver with a poor aerial.

5.2 Impedance

An aerial, like an electrical circuit, has a certain impedance. Each aerial has a certain inductance and capacitance and for any given frequency the aerial will have a certain impedance. At resonance the inductance and capacitance cancel one another out, leaving a purely resistive element. This is known as the **radiation resistance**, and varies from one type of aerial to another.

5.3 Gain and directivity

One important characteristic of an aerial is the way in which it is more sensitive to signals in one direction than another. This is called the **directivity** of the aerial.

To explain some of the features of directivity it is easier to visualize the operation of the aerial when it is transmitting. The aerial performs in the same manner when it is receiving.

Power delivered to the aerial is radiated in a variety of directions. The aerial design may be altered so that it radiates more in one direction than another. As the same amount of power is radiated, it means that more power is radiated in one direction than before. In other words it appears to have a certain amount of **gain** over the original design.

In order to see the directional pattern of an aerial a **polar diagram** is plotted. This is a plot of the signal strength around the aerial, the distance of the line away from the aerial indicating the relative strength. A simple dipole aerial may have a pattern like that shown in Figure 5.1. From this it can be seen that the maximum radiation for transmission, and hence the maximum sensitivity to received signals, occurs when the signal is at right angles to the aerial.

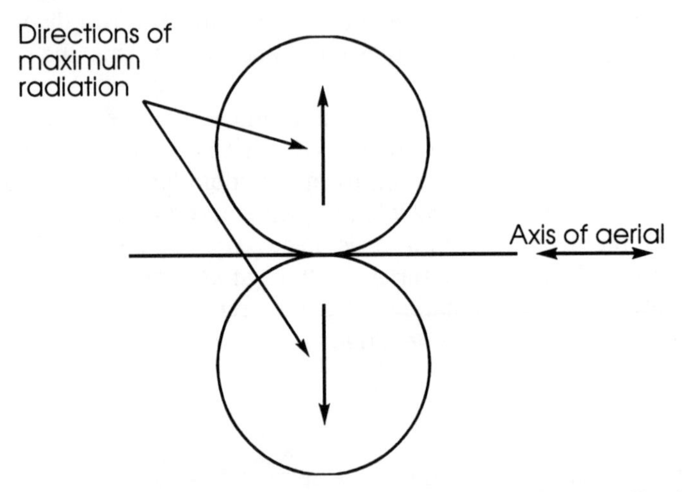

Figure 5.1 *Polar diagram of a half wave dipole*

When an aerial is designed to be directive, its polar diagram may look more like that shown in Figure 5.2. In this diagram it can be seen that the aerial radiates far more signal in one direction than another. Because it 'beams' the power in a particular direction, aerials of this type are often called **beams**.

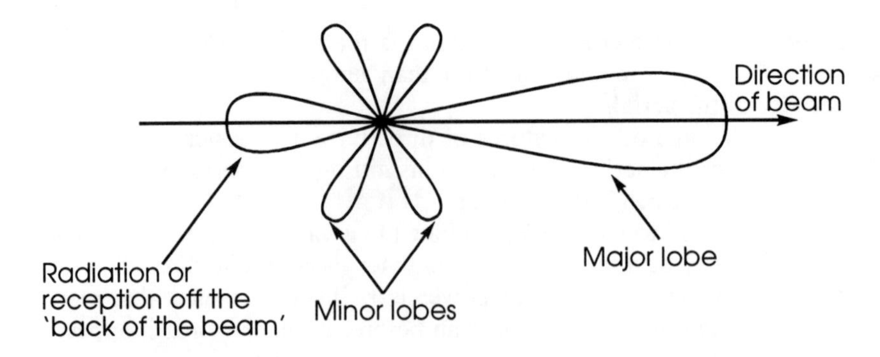

Figure 5.2 *Polar diagram of a directive aerial or beam*

The region of maximum radiation is called the **major lobe**. However, there are other areas around the aerial where there are significant levels of radiation. These are called **minor lobes**. They are always present to some extent and generally the largest is in the opposite direction to the main lobe. Sometimes the front to back ratio is of interest and a front to back ratio is quoted and expressed in decibels. This is simply the ratio of the signals in both directions expressed in decibels. It is of particular interest in situations when there are likely to be significant levels of interference in the opposite direct to a wanted signal.

One of the major design parameters of any beam aerial is its gain. This has to be compared to another aerial. The most common aerial used for comparisons is called a dipole. The gain is simply the ratio of the signal from the beam aerial compared to the dipole expressed in decibels.

Sometimes another type of aerial may be used. Called an **isotropic source**, it is an imaginary aerial that radiates equally in all directions. It can be calculated that a dipole has a gain of 2.1 dB over an isotropic source and therefore when any aerials are quoted against an isotropic source the gain is measured against a dipole and then 2.1 dB is added.

It is necessary to specify what the gain of the aerial is being compared against. The gain of an aerial over a dipole is quoted as a certain number of dBd (dB gain over a dipole). Similarly figures against an isotropic source are quoted as dBi (dB gain over an isotropic source).

5.4 Angle of radiation

An aerial like a horizontal dipole will radiate power at all angles relative to the earth, ignoring the effects of the ground nearby. This means that some power will travel vertically upwards towards the ionosphere. If this can be reflected back to the ground it will be received by nearby stations. Other power will travel almost horizontally along the ground reaching the ionosphere and being reflected back to the ground much further away. Whether the aerial is horizontal or vertical like that shown in the diagram, the angle of radiation is very important.

Distant signals usually arrive at the aerial at a low angle. This means that if distant stations are being sought it is necessary to use an aerial with a low angle of radiation. Having an aerial like this also helps reduce interference from less distant stations as they are usually received at a much higher angle. Vertical aerials generally have a low angle of radiation and are much less sensitive at higher angles, making them ideal for long distance work.

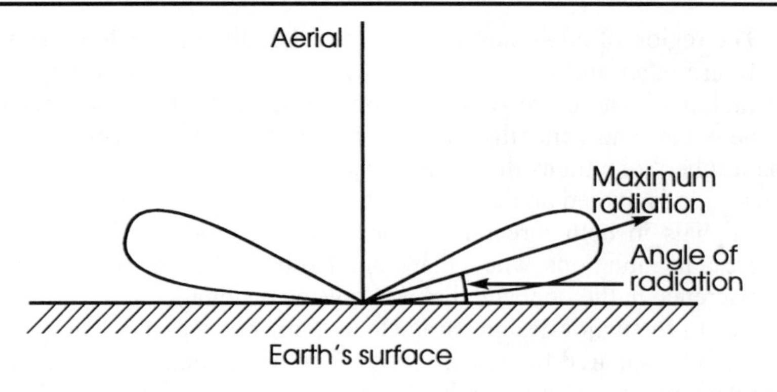

Figure 5.3 *Angle of radiation*

5.5 Height

The height of an aerial is very important. If it is installed close to the ground then it will be shielded by nearby objects, reducing the overall signal levels, especially the low angle long distance signals. Raising the height of the aerial will not only enable the aerial to see over these objects, but it also increases the distance to the horizon, again increasing the ability to receive the long distance low angle signals.

It is difficult to estimate the signal level increase obtained by raising the height of an aerial. Less distant stations which are being received at a higher angle will be affected less than longer distance stations at a lower angle. However, any real increase in height will give a useful degree of improvement.

5.6 Aerial system

It is not possible to consider the aerial elements in isolation. It is necessary to look at the whole aerial system and optimize the operation of the whole system. The aerial system can be split down into three main parts: the aerial elements themselves, the feeder and the matching unit providing the correct match between the aerial and the feeder. This is used to transfer the power from the place where the aerial itself is located into the receiver. It is particularly important because the requirements for optimum performance rarely coincide with the ideal location for the receiver.

5.6.1 Feeder

The purpose of a feeder is to carry radio frequency signals from one point to another with the minimum amount of signal loss. Because radio frequency signals are being carried, ordinary wire like that used for carrying mains power is unsuitable.

There are two main types of feeder used for short wave signals. One is called **coax** or **coaxial feeder**. This type of cable is used for connecting television aerials to the set. It consists of two concentric conductors, spaced apart from one another by an insulating dielectric as shown in Figure 5.4.

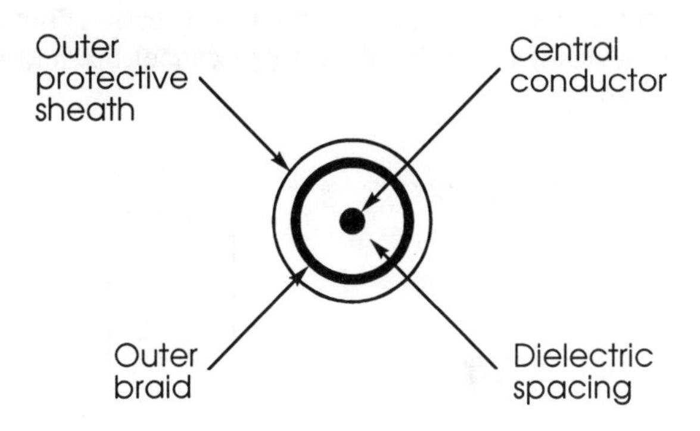

Figure 5.4 *Cross-section through a piece of coaxial cable*

The inner conductor may be single or multi-stranded, but the outer one consists of a braid as shown in the diagram. The outside of the cable is covered by a protective sheath to prevent moisture ingress, as well as providing some mechanical protection.

Coax is the most commonly used type of feeder. It has a number of advantages, including the fact that it is not affected by nearby objects. This means that it can be run almost anywhere without any ill effects.

There are a number of different types of coax. The main distinguishing feature is the impedance. For most television and domestic hi-fi aerials, 75 ohms has been adopted as standard. For amateur radio, Citizen's Band and commercial applications, 50 ohms is used as the standard. When buying coax for a short wave listening installation it is necessary to obtain the correct type and this is normally 50 ohms. The 50 ohm types cannot normally be bought in domestic electrical shops as they generally stock only the 75 ohm type for television and hi-fi installations. An electronics shop or amateur radio dealer is the most likely source.

The other main specification for coax cable is its loss. The thinner cables generally have a higher loss than the thicker ones. It is also frequency dependent, rising with frequency. The loss is generally quoted as a loss over a given length, normally 10 metres and at a certain frequency. Fortunately the loss of most cables is comparatively small for frequencies below 30 MHz, unless long lengths are required. The choice is generally a balance between cost and performance as the thick low loss types can be very expensive.

Coax is an unbalanced cable. In other words, the outer braid is connected to earth. In some instances balanced feeders need to be used. Often called **twin** or **open wire feeder,** it is not nearly as widely used as coax.

A balanced feeder consists of two wires spaced evenly apart. The currents flowing in both wires run in opposite directions but are equal in magnitude. As a result the fields from them cancel out and no power is radiated or picked up.

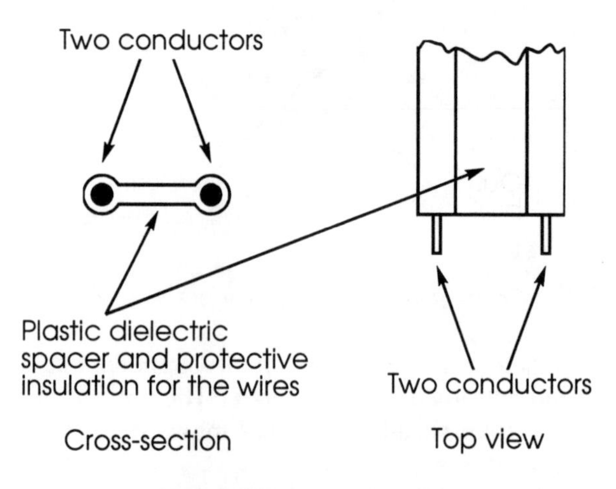

Figure 5.5 *Twin feeder*

Several versions of this can be bought ready made. The best type uses a black plastic dielectric and insulation. Also the plastic between the wires is not continuous and has oval shaped spaces to improve the performance. The clear plastic type is not nearly as good because the plastic dielectric absorbs water and the level of loss rises.

An alternative to the factory-made twin feeder is open wire feeder which can be easily made for very little cost. It is made by taking two wires and placing spacers at intervals as shown in Figure 5.6. These spacers should be placed about 30 to 50 centimetres apart and can be made from almost any suitable insulator. Small lengths of plastic have been used and holes drilled for the wires to pass through. The spacing is not critical for many applications and can be made around 6 to 10 centimetres. This type of feeder is very cheap to make up and gives a very low level of loss, provided it is kept clear of nearby objects.

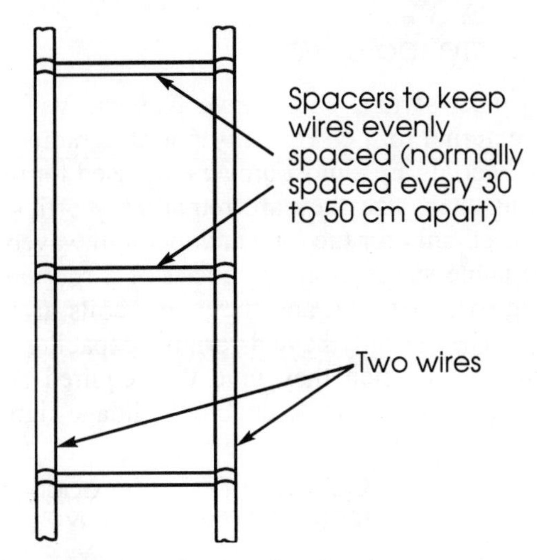

Figure 5.6 *Open wire feeder*

The major problem with this type of feeder is that it is unbalanced by nearby objects and should not be run through the house. This makes it far less convenient to use than coax.

5.6.2 Matching unit

In order that the maximum amount of signal is transferred from the aerial to the feeder, the impedances must be the same. Unfortunately the impedance of the aerial will change with frequency, so that an aerial designed to give a good match at one frequency may not be so good at another.

To ensure that the best match is obtained, an aerial tuning or matching unit can be used. There are a number of designs for these units which contain coils and variable capacitors so that the aerial can be tuned to resonance and give a good match to the feeder. In this way the aerial can be made to operate at its greatest efficiency over a band of frequencies.

5.7 Types of aerial

A wide variety of aerials is available for the short wave listener. The choice of aerial will depend on a number of factors, but in most cases one capable of covering a wide range of bands is needed. If there is particular interest in one band, a specific aerial cut for that band may be an ideal choice. However, the choice of the best aerial will depend upon the type of listening envisaged, the location and space available.

5.7.1 Ferrite rod aerials

Although most serious short wave listeners will want to use an aerial which is external to the set, many portable radios contain internal ones. Ferrite rod aerials are almost universally used for this type of application. Considering their size they are remarkably efficient, although they are normally used only for the long and medium wavebands.

As the name suggests, they consist of a rod made of ferrite, an iron-based magnetic material, and they have coils around them as shown in Figure 5.7. They usually have a variable capacitor connected across them for tuning and as such they give the required selectivity to the radio frequency stages of the set to enable the image signal to be reduced.

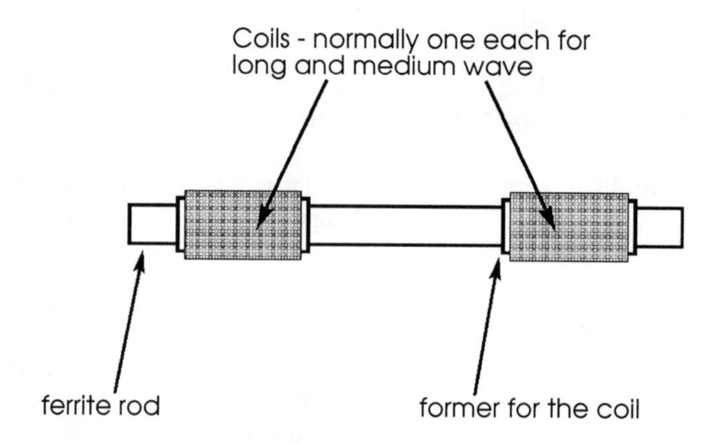

Figure 5.7 *A ferrite rod aerial*

A ferrite rod aerial is directive. The minimum pickup is in the line of the rod, and the maximum at right angles. This may appear to be a disadvantage for normal listening. However, when trying to pick up distant broadcast stations a local station can be reduced in strength to let the required station through by suitably positioning the set.

Whilst these aerials are very convenient for many types of listening, they cannot compete with many types of larger aerial. They are also not suitable for use at frequencies much above 2 MHz. The loss in the ferrite increases with frequency and reduces the efficiency of the aerial.

5.7.2 Telescopic aerials

Many portable or 'World Band' radios have their own telescopic aerials. Those used for short wave sets usually extend further than those used only for VHF FM. Even though they are quite long, they are not sufficient for serious short wave listening. However, they do provide sufficient performance for most sets to pick up a good number of the stronger broadcast signals.

It is also worth bearing in mind that many sets which have these telescopic aerials may not be capable of handling the much larger signal levels provided by a good aerial. The radio frequency circuits may become overloaded, leading to the generation of unwanted spurious signals like intermodulation products.

5.7.3 End fed wire

One of the most popular aerials for listeners is a single wire aerial. Although they are normally called long wires, the more correct term for the majority is an **end fed wire**. A true long wire is many wavelengths long and may be 100 metres or more, dependent upon the wavelength in use.

The advantage of an end fed wire is that it is easy to install and can be used over a wide range of frequencies, especially if a tuning unit is employed.

The aerial simply consists of a length of wire, reasonably long and as high as possible. For a typical listening installation a length of around 10 metres or so is suitable, although if it is longer or shorter this is not a problem. However, if a particular band is favoured, a length of a quarter of a wavelength is ideal.

A typical wire installation is shown in Figure 5.8. The wire can be strung between almost any two points. Convenient points are generally the house and a pole at the remote end of the garden, or possibly a tree.

If a tree is used, then account must be taken of its movement in the wind. If this is not done then the wire will break when the tree moves in the wind. A system like that shown in Figure 5.9 can be used for this. The tension in the aerial is governed by the weight and this should be chosen to give enough tension to keep the wire reasonably taut without applying so much that it places undue strain on any point of the aerial. Ideally the minimum to keep the wire from sagging too much should be chosen. Other points to be remembered are that the weight must be positioned so that there is no chance of it hitting anyone walking by the tree. Also a loop of rope can be taken from the weight back to the tree as shown. This helps it not to swing around too much in the wind and become caught. However sufficient slack must be allowed for it not to become taut even in the windiest of conditions.

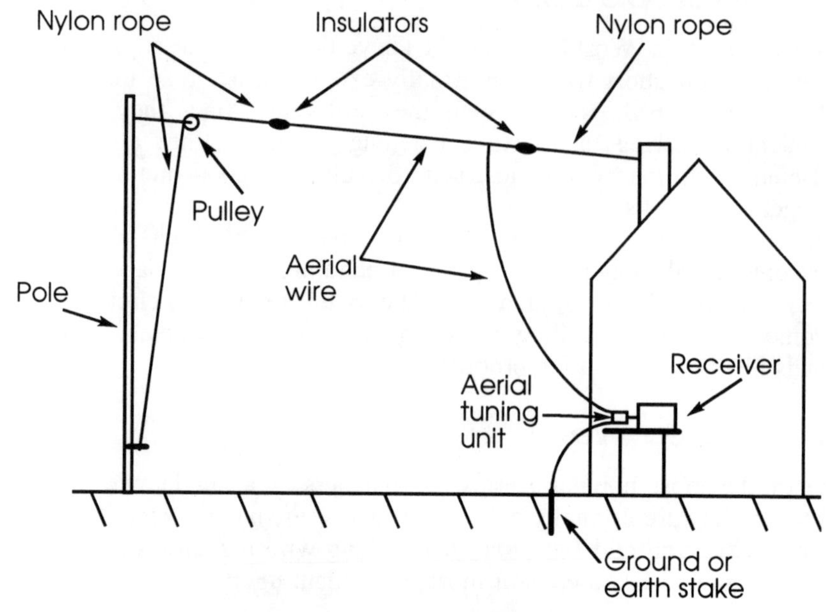

Figure 5.8 *An end fed wire aerial*

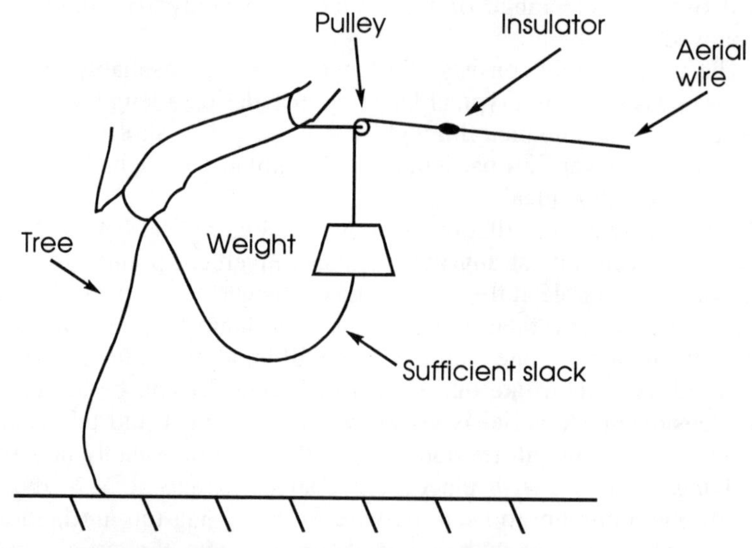

Figure 5.9 *Anchoring an aerial to a tree*

Insulators are normally used at either end of the wire. These are used to prevent the rope holding the wire up from affecting the performance of the aerial. Whilst polypropylene rope is often used, the insulator provides a convenient method of interfacing the rope and wire. It also helps reduce the effects of any moisture on the rope.

There are two main types of insulator. The most common type is known as an egg insulator. It gains its name because of the shape as seen in Figure 5.10. This type is easy to use and safe. If it breaks for any reason the rope and wire will still hold because the rope and wire are still looped through one another. Today most of these insulators are made from plastic, although there are still many of the ceramic varieties available.

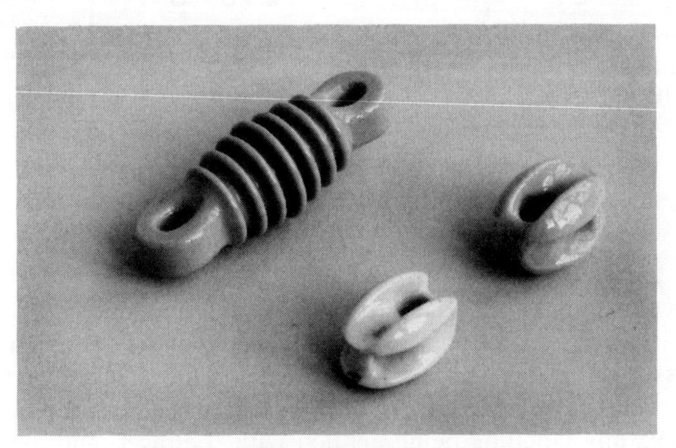

Figure 5.10 *Aerial insulators*

The other type of insulator is also shown in the photograph. These are more expensive and used less frequently for amateur installations. The insulators are generally made of plastic although older ones may be obtained occasionally and they are made of either glass or glazed ceramic. Any of these insulators will provide excellent insulation, because the ribbed effect ensures the maximum surface distance so that the highest resistance is obtained.

The wire from the main section of the aerial can be brought into the room via a suitable route. However, it is advisable to keep the lead away from any mains wiring as it can still pick up signals and may pick up any mains-borne interference.

Once in the room, the aerial needs to be connected to an aerial tuning unit. This need only be a simple unit such as that shown in Figure 5.11. A simple unit can be made up quite easily, requiring only a single tuning capacitor, coil and switch as the main components.

The coil can present some problems. The former should be an inch or two in diameter; you may be able to obtain a ceramic coil former. However, these are expensive to buy new and unless they are bought as surplus they are not normally worth the expense for a listening station. A suitable alternative can be made out of 1½ inch white plastic waste pipe. Grooves are needed to hold the wire in place. These can be cut using a lathe or any other suitable method. It is best to ensure that one turn is spaced from the other by at least the diameter of the wire. This enables taps to be soldered onto the wire as shown in the photograph. The wire should be reasonably thick, at least 20 SWG.

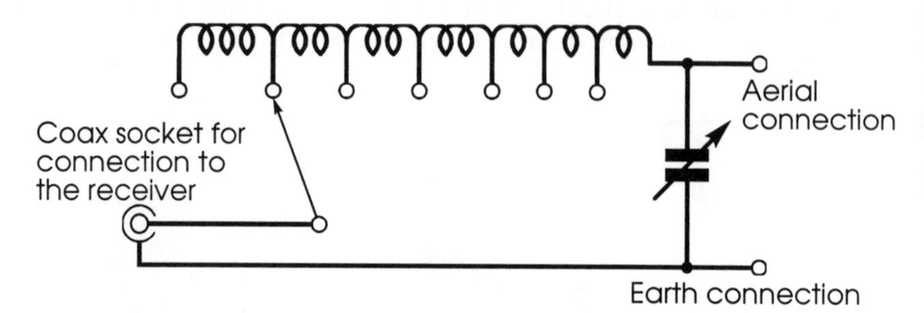

Figure 5.11 *Circuit of a simple aerial tuning unit*

The connections for the aerial and earth can be made using banana or 4 mm sockets. The connection to the receiver will be via a coaxial cable and possibly the most convenient connector for this connection is an SO239 socket. The box can be conveniently made from a ready painted metal box, any labelling being applied with 'Letraset' and then covered with clear adhesive film, like that used for covering books. This provides a sufficiently tough cover, whilst maintaining a professional looking finish.

Figure 5.12 *View of the inside of a home-made aerial tuning unit*

5.7.4 Earth connection

Aerials like end fed wires require a good earth system if they are to perform to their best. Whilst most electrical equipment is connected to the mains earth this is very noisy from an radio viewpoint, although vital for safety.

A good radio earth can be made quite easily. It can simply be a rod driven into the ground. In fact it is possible to obtain rods specially for this purpose. They have copper on the outside to give good conductivity and steel on the inside to give the strength for it to be driven into the ground.

The ideal place for an earth is where the ground is relatively wet, as this increases the conductivity. However the lead from the receiver to the earth must be as short as possible. It is also good practice to make the wire from the earth rod to the receiver reasonably thick to ensure a low level of extra impedance.

The earth connection does not have to be made from rod. Some listeners have buried old water tanks and found that these make very successful earths as their large surface area makes a low resistance connection.

5.7.5 Dipole

The dipole is one of the most popular forms of aerial. It is widely used on its own or as the basis of other directive aerials.

The most common form of the dipole is the **half wave dipole**. From the diagram it can be seen that it consists of two quarter wavelength sections, each connected to the feeder. This type of dipole gives a good match to 50 ohm coaxial cable, although it may also be fed with twin feeder.

It is easy to determine the length of a dipole. The exact length is not quite the same as a half wavelength in free space, but slightly shorter. This is due to a number of effects including one referred to as the **end effect**. To calculate the length required for any given frequency one of the following formulae can be used:

$$\text{length (metres)} = \frac{148}{\text{frequency (MHz)}}$$

$$\text{length (feet)} = \frac{480}{\text{frequency (MHz)}}$$

Once the length has been calculated, it is best to cut the wire slightly longer and trim it for optimum performance once it is in situ. This is done because it is more difficult to add wire to the aerial in a satisfactory manner than cut it off.

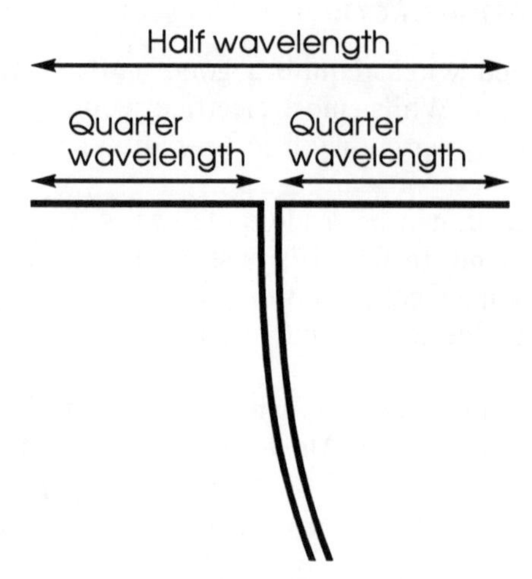

Figure 5.13 *A dipole aerial*

A dipole is a balanced aerial. In other words, neither of the two sections is connected to earth. If coaxial cable is to be used as the feeder, then a **balun** should be used as shown in Figure 5.14. A balun is an RF transformer which can be used to convert from balanced to an unbalanced format. If the aerial impedance is the same as the feeder, the transformer has a 1:1 ratio and the impedance stays the same. However, the balun can be used to transform the impedance by having a suitable turns ratio.

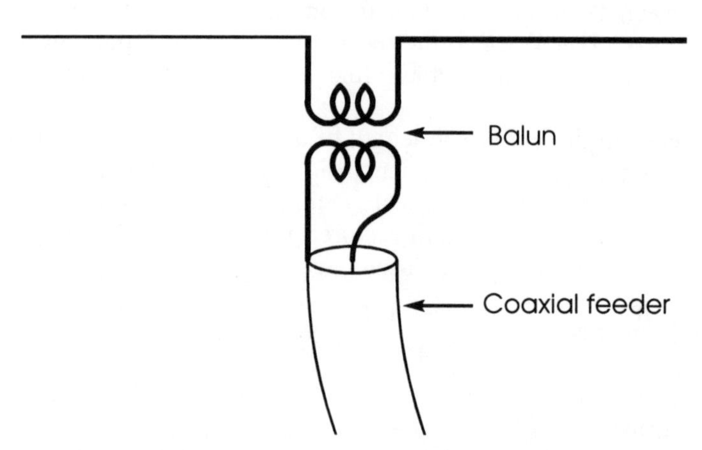

Figure 5.14 *Using a balun with a dipole aerial*

Often balanced aerials will be fed with coax without the use of the balun. Whilst the aerial will still work if it fed in this fashion, unwanted currents will flow down the outside of the feeder, allowing signals to be picked up along its length. The impedance of the aerial may be affected as may its radiation pattern.

The installation of a dipole follows many of the same rules as an end fed wire. The main difference is that the aerial is centre fed, and this requires a feeder at the centre of the aerial as shown in Figure 5.15. Ideally this should be taken away at right angles to the axis of the aerial so that the feeder does not affect the operation of the aerial itself.

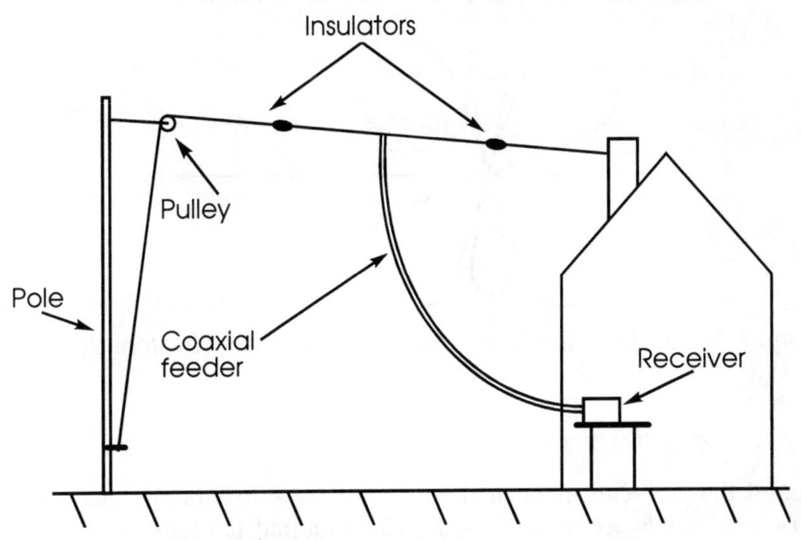

Figure 5.15 *A dipole aerial installation*

To make a satisfactory join at the centre of the aerial, special centre pieces are available. These enable the coax and the two wires to be held without the strain being taken by the joints between the aerial and coax. In view of this it is wise to use a centre piece as it increases the mechanical reliability.

If the dipole is mounted outside, the coax will be open to the elements. It is very important to ensure that water does not enter the end of the coax. If this happens then moisture can penetrate a long way into the feeder, increasing the level of loss very considerably. It is therefore wise to seal the end to prevent this happening. Self-amalgamating tape is best but for short wave installations it is quite adequate to use one of the silicon rubber bath sealants or even a contact adhesive like 'Evostick'. This can be applied liberally around the open end.

Dipoles do not have to be a half wavelength long. They can be longer. Often three half wavelength aerials are used. These still provide a good match to coax as they are fed at the current maximum point. Similarly five half wavelength aerials can be used. The main advantage of using a long aerial is that it can be used on a number of frequencies. For example an aerial trimmed for the 7 MHz amateur band will also be resonant at 21 MHz where there is another amateur band.

The major difference is in the directional properties of the aerial. The half wave aerial has its maximum radiation or pickup at right angles to the axis of the aerial. The three half wavelength aerial has a different radiation pattern as shown in Figure 5.16. Similarly longer aerials have the major lobes moving more towards the axis of the aerial.

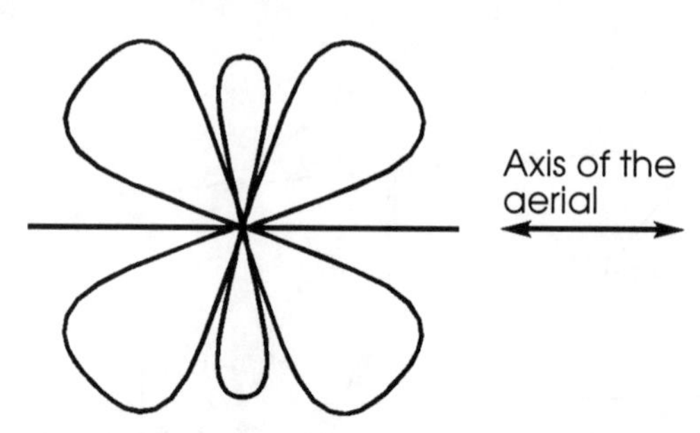

Figure 5.16 *Directional pattern of a three half wavelength aerial*

5.7.6 Trap dipole

One of the problems associated with short wave listening is that a wide variety of bands are usually used. Whilst an end fed wire with a tuner offers an ideal solution, many people will want to use a dipole. There are a number of designs of dipole which can offer operation on a number of bands. One method of achieving this uses **traps** or resonant circuits placed in the aerial to isolate various sections. The trap also has the effect of adding inductance to the aerial or loading it. This results in its length being reduced on the lower bands of operation.

A typical two band dipole is shown in Figure 5.17. The section of the aerial with a length l_1 between the two traps forms a traditional half wave dipole for the highest frequency band. The traps are tuned to this frequency, isolating the remainder of the aerial at this frequency. For the lower frequency, the aerial has a physical length l_2 and again performs as an electrical half wavelength. However, the inductance in the traps has the effect of reducing the physical length required. This can be an added advantage in using traps in a multiband aerial.

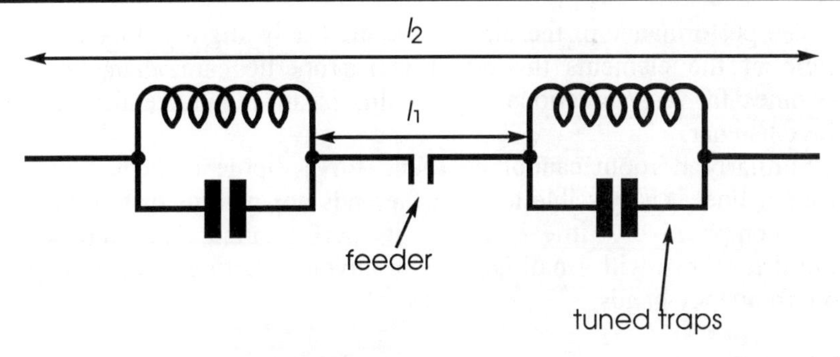

Figure 5.17 *A trap dipole*

5.7.7 Inverted V

One of the major problems with a dipole is that it requires two supports and is fed in the middle. It is found that the maximum pickup or radiation of a signal usually occurs in the middle. It would seem most sensible to make sure that the aerial is highest at this point.

Because of this the inverted V aerial is often used. As shown in Figure 5.18, this aerial is in the form of an inverted V with the centre at the highest point. Often the ends are anchored to suitable low points, leaving only one high fixing for the centre. However, care must be taken to ensure that the low end points do not become a hazard that people can walk into.

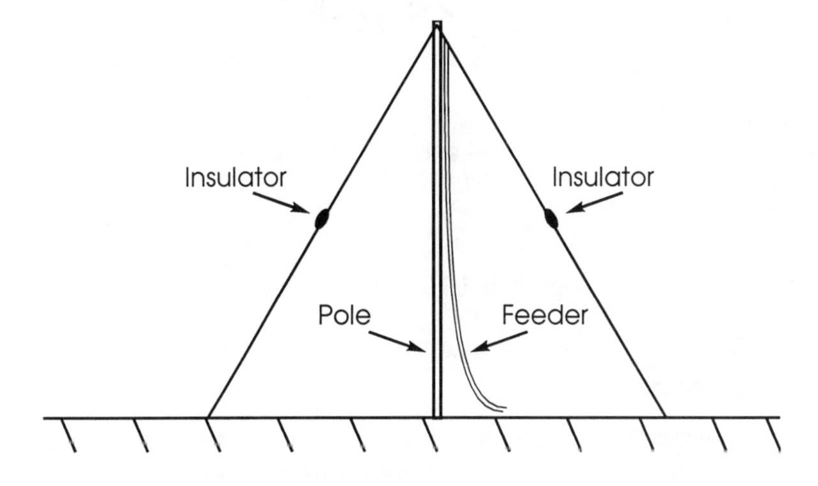

Figure 5.18 *An inverted V dipole*

The performance of the dipole is altered only slightly. In view of the angle of the elements the directional properties are changed and it becomes far less directional. Often this is an advantage for the short wave listener.

Similarly if room cannot be found for a dipole to be erected in a straight line, it is possible to bend the ends downwards or to either side with comparatively little effect on its performance. This can be very useful for those with small gardens or anyone wanting to operate on the low frequency bands.

5.7.8 Doublet

It is not always necessary to feed a dipole with coax. If it is fed with open wire feeder, then it is possible to locate the aerial tuning unit at the end of the open wire feeder as shown in Fig 5.19. This means that the aerial can be used over a range of frequencies. Provided that the length is over a half wavelength at the lowest frequency of operation it will perform efficiently. Even when it is shorter than this it will still be satisfactory for listening, although there will be a reduction in performance.

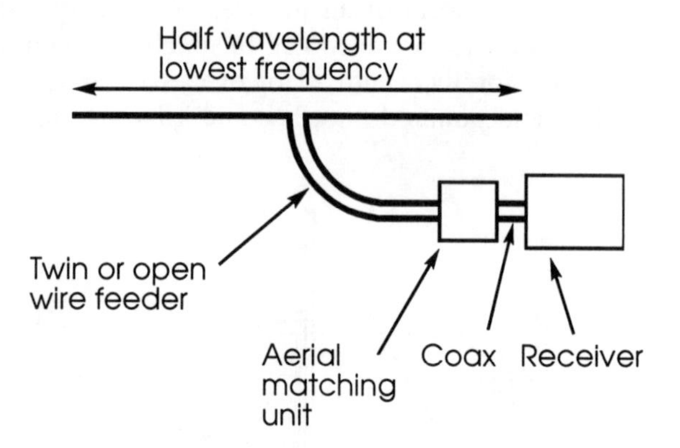

Figure 5.19 *A doublet aerial*

5.7.9 Beam

Many transmitting stations use beams. These aerials can give several decibels gain which comes in very useful when making the most of the available transmitted power. Whilst a beam can be used to good effect for receiving, few listeners go to the expense of putting up one of these aerials.

There are a number of different types of beam which can be built or bought. However, the most common by far is called a **yagi**. It is the same basic design as most television aerials, although the size of the elements is much greater because of the frequencies in use. Also there are fewer elements for the same reason.

The aerial is based around the dipole. Additional parasitic elements are placed either side of this and they have the effect of changing the radiation pattern of the dipole. The parasitic elements are made longer or shorter than the main or driven element. By making the parasitic element longer it acts as a reflector. This is placed behind the main element to reinforce the signal. By making the parasitic element shorter it acts as a director and is placed in front of the main element. Normally only one reflector is used because additional reflectors do not give any noticeable improvement in performance. Adding additional directors does give additional gain and often more than one director is used. In television aerials a large number of directors are commonly used. For the short wave bands the aerials become too large if a number of directors are used. Most beams only use one director and even for the highest frequencies on the short wave bands it is unlikely that more than four are used because this makes the aerial extremely large.

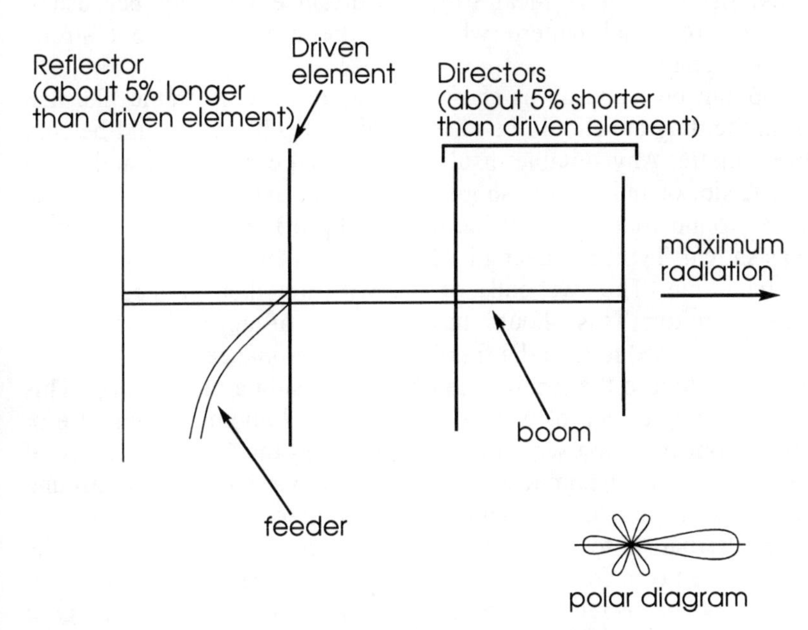

Figure 5.20 *A yagi aerial*

From the diagram it can be seen that the maximum gain will be in the direction of the directors, although signals can still be heard off the back of the beam. Often a front to back ratio is specified. This is the ratio of the sensitivity of the aerial in the forward or main direction to its sensitivity off the back of the beam. A good front to back ratio is desirable when there are large numbers of strong signals in the opposite direction to the wanted one. Many amateur stations in the USA require a good front to back ratio because of the levels of interference from Statesside stations. Apart from the lobe at the back of the aerial there are a number of other minor lobes. These are usually tolerated and are of less significance than the one at the back of the beam.

Normally a beam aerial can operate on only one band. As they can be very large, it is not usually possible to put more than one of these aerials up. This has led to the development of trapped aerials using the same system as the dipole. These trapped beams are often used by radio amateurs. The maximum number of bands they can be used for is normally three.

5.7.10 Loop aerial

A loop aerial provides an alternative to the ferrite rod aerial, especially for medium wave listening. Although it is not widely used for normal broadcast listening, it is ideal for long distance listening because it exhibits a directional pattern which can be used to cut out strong interfering signals.

A loop can be made up as shown in Figure 5.21. A frame like that shown in the diagram should be made up first. Each side should be very roughly 1 metre. Any suitable insulating material can be used and wood is ideal. A slot or indentation should be made at the end of each arm so that wire wound round the outside cannot slip off. Wire is then wound round as shown. Typically a length of around 40 metres should be wound round the frame. The two ends of the wire are then connected to a variable capacitor. This should have a range of approximately 0 to 500 pF and it is used to tune the frame aerial to resonance.

The connection to the receiver can be made using a pickup loop. This consists of a single loop of wire, normally placed about a centimetre or two away from the main winding. This preserves the Q or selectivity of the loop. Normally the impedance of loops is very low, often around 1 ohm and a low impedance input may be needed for this.

As an alternative to winding a pickup loop, the connection to the receiver can be taken from either side of the tuning capacitor. This is not as satisfactory because the loading effect of the receiver reduces the Q of the aerial and the fact that one end is grounded destroys the balance. However in real terms the performance may still be quite satisfactory.

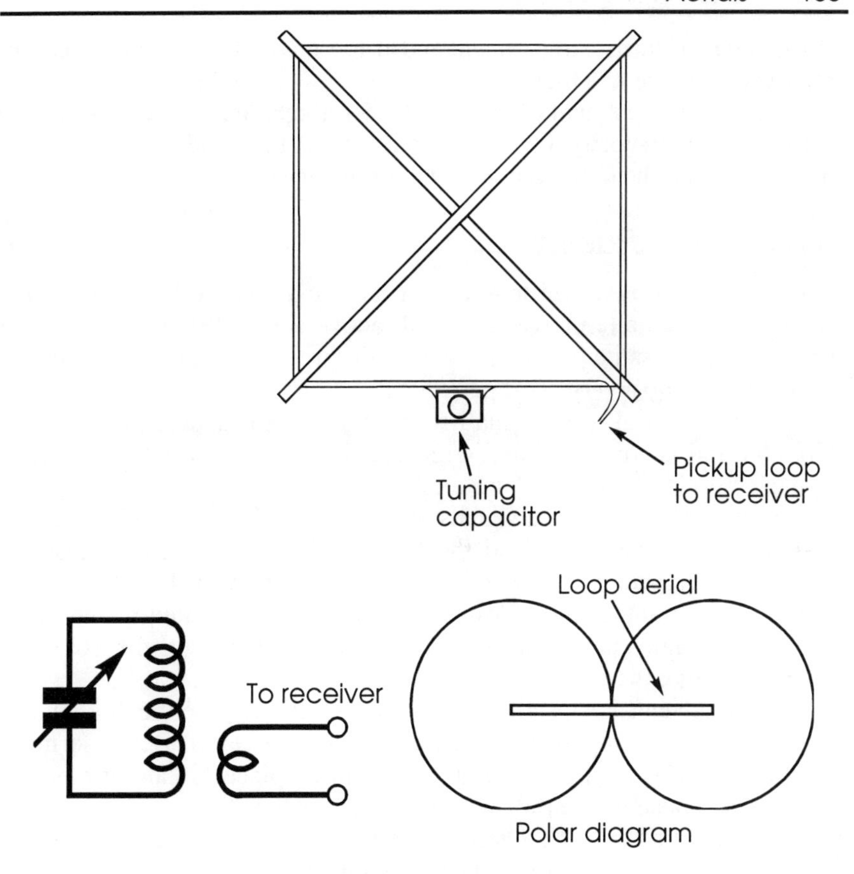

Figure 5.21 *A loop aerial*

If no additional preamplifier is to be used, the connection can be taken to the two aerial input terminals on the receiver. If a coaxial connector and the additional one-turn pickup loop is used then either connection can be taken to the centre and earth on the coaxial connector. If no pickup loop is employed then the wire connected to the earth terminal on the capacitor (the one connected to the moving vanes) should be connected to the outer connection on the coaxial connector.

The loop aerial may need a little adjustment once it is built. It should be possible to tune the capacitor to give maximum signal strength at both the top and bottom of the required frequency range. If it cannot achieve this over the top end of the band it may be necessary to take some turns off the loop. This should be done one turn at a time so that the correct number are taken off. Alternatively, if it cannot cover the bottom end of the band, additional turns may be needed. In this case a judgement about the number of turns can be made and if too many are added they can be removed one turn at a time.

After each adjustment check that the aerial operates at the other end of the range satisfactorily. Otherwise, one end of the band will be made to be correct, and the other may no longer tune properly.

5.7.11 Active aerial

It is not always practical to set up a large external aerial for the short wave bands. An alternative is a small active aerial. Whilst these aerials cannot give the same performance as their larger traditional counterparts, they can provide a very good performance.

An active aerial uses a small receiving element and some electronic circuitry and is connected to the receiver by a length of ordinary feeder. The electronics consists of an amplifier. This provides two functions. The first is to amplify the signals and the second is to provide a better match between the aerial and the feeder. This is required because the aerial is very short when compared to a quarter wavelength and will present a high impedance. Power for the circuitry located in the aerial itself is normally provided along the inner conductor in the coax, the outer providing the return.

The aerial can be installed in a suitable location, either indoors or outside depending upon the specifications of the aerial. Being so much smaller than other types of aerial, they can be placed in an attic or just outside the house where they will hardly be visible.

The design of an active aerial is not easy. Most of these aerials are capable of operating over a wide bandwidth, covering possibly the whole of the short wave spectrum. This means that there are likely to be a number of very strong signals. The amplifier must be designed so that it does not overload, otherwise it will generate spurious signals. As a result good active aerials can be expensive, although some cheaper kits are available which give good value for money.

5.8 Aerial siting

Choosing the correct site for an aerial can enhance its performance quite considerably. To ensure that the best performance is obtained a few pointers should be observed.

First of all, the aerial should be kept away from any obstructions which may screen it from the signals it is trying to pick up, particularly if the obstruction is metallic. However even trees and wooden poles will have an effect, especially when they are wet.

The aerial should also be as high as possible, as mentioned before. This too has a significant effect.

If at all possible, it is worth keeping the aerial away from the house. Large amounts of electrical noise are generated by many pieces of electrical equipment found around the home. The line scan circuitry in televisions creates a rasping signal every few kilohertz which can be heard up to frequencies of 10 MHz and more, although some of the newer sets are suppressed much better than the old ones. Computers, fluorescent lights, vacuum cleaners, electric drills and a host of other devices all produce noise which can be picked up. By keeping the aerial away from these as much as possible, reception can be improved.

5.9 Safety

Safety is an important factor in aerial design and installation and it should be kept in mind when any new aerial is being erected or when maintaining an existing one. External aerials have to withstand the rigours of the weather. Strong winds can place high levels of loading on the various elements in the aerials. Accordingly, the proper fixings must always be used. Besides this, the corrosive effects of the weather can soon render otherwise strong fastenings much weaker than normal.

Also, when the aerial is installed, it is very important to ensure that there is no chance of it falling across any power lines. This has happened in the past and it can be very dangerous. People have been killed when this has occurred.

Great care must also be taken when installing the aerial. Work often has to be undertaken on ladders; again there have been casualties in the past. However, if the right precautions are taken, the chances of any mishaps can be reduced and the rewards of installing a better aerial can benefit one's enjoyment of the hobby by bringing in better and more distant signals.

5.10 Lightning

One of the major concerns of radio amateurs and short wave listeners is the effects of lightning. The energy stored in a lightning strike is phenomenal. Fortunately direct strikes occur comparatively infrequently in the UK, although in other parts of the world where electrical storms occur more often there is obviously a greater chance of damage. However, when aerials are erected as high as possible this does make them more susceptible to strikes. It does not require a direct strike to cause damage to equipment. Even strikes some distance away can induce very large voltages in an aerial. Sometimes sparks have been seen to jump across from disconnected aerials to an earth point. If a receiver is connected to an aerial under these conditions there is a very real chance of damage resulting, apart from the possibility of electric shocks.

Commercial radio stations have to be capable of operating through lightning storms and their aerials have to withstand direct strikes. For the listener, aerials are not so large and it is possible to close down the station when there is a possibility of a storm. Accordingly when a storm is likely it is best to disconnect all aerials from the equipment and ground them outside the house or building. This will prevent static build-up. It may also be wise to lower the aerial, particularly if it is very high. If the aerial has a rotator, the cables from this should also be disconnected.

A further precaution is to disconnect the receiver from the mains supply. When lightning strikes power lines it is possible for spikes and surges to travel along the lines and enter the home. Although it is at a much reduced level, it can nevertheless be sufficient to cause damage in some instances.

6 Ancillary equipment

Apart from the basic receiver there are many items which can be added on to the set to improve the performance or add additional facilities. Items like aerial tuning units enable the maximum amount of energy to be transferred from the aerial into the receiver, thereby increasing the effectiveness of the station. Other units are available which allow digital data to be decoded and information displayed either on paper or a screen. This opens up a whole new realm of short wave listening.

6.1 Aerial tuning units

One piece of equipment which will be of use in almost any set-up is an aerial tuning unit (ATU). End fed wires are very popular aerials for listening because of their simplicity and flexibility. However, to ensure that they give their best performance an aerial tuning unit is required, like the one described in Chapter 5. Apart from their use with end fed wires, aerial tuning units can be used with many other types of aerial as well.

The tuning unit ensures that the aerial impedance matches the feeder or the receiver input impedance. In this way the maximum amount of signal is transferred from the aerial into the receiver.

Another advantage of an ATU is that it is a tuned circuit and will allow signals through on some frequencies and reject or attenuate others. This means that once it is tuned to the required frequency or band it will tend to reject others. As the out of band signals are reduced in strength, image rejection and other similar parameters may be improved.

When buying or building a tuning unit, it is necessary to ensure that it is suitable for the types of aerial to be used. Some are only suited to operation with coaxial cable, whereas others may only be used with end fed wires. Some are not suitable for use with balanced feeder. Most tuners accept coaxial feeders and end fed wires, but a limited number can accept balanced feeders.

A tuning unit is normally connected into the receiving system as shown in Figure 6.1. With an end fed wire the aerial itself is connected directly into the tuning unit. There should be a specific connection for this and also a connection for the earth. With a balanced feeder there should be two connections, neither of which is earthed.

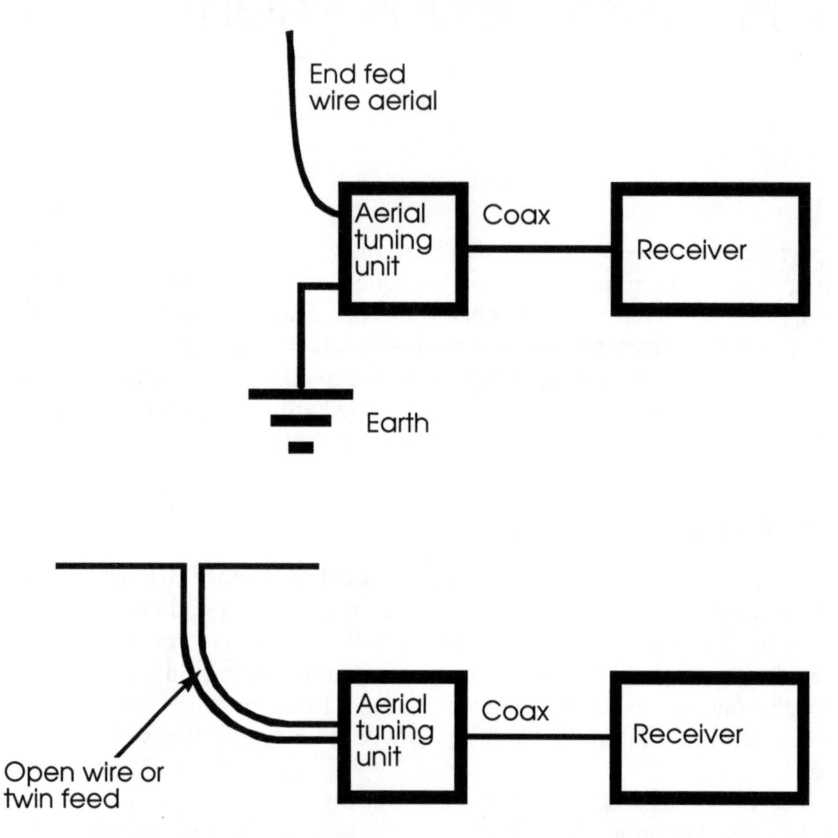

Figure 6.1 *Receiving systems with an aerial tuning unit*

On the tuning unit there will be a number of controls. Usually there is one to alter the inductance. This may be a switch, or in some cases it may be a 'roller coaster'. This is a continuously variable inductor. In addition to this there will be one or two variable capacitors. These should be adjusted to give the maximum signal strength. Unfortunately this is often easier said than done. First it is best to tune the set to a station where the strength is constant, or nearly constant. Then the controls can be adjusted to give the optimum setting. Once this has been done it is wise to note the settings and the frequency so that the next time the band is used, it is much easier to set the unit.

6.2 Preselectors

Preselectors are not nearly as common as they used to be some years ago. They consist of a radio frequency amplifier and some RF tuning. Many old sets, particularly some of the valve receivers which were on the market some years ago, were not very sensitive. The amplification provided by the preselector enabled the sensitivity to be considerably improved and the radio frequency tuning prevented parameters like the image and other spurious responses from becoming degraded.

Most receivers today have very good sensitivity and adding a preselector with significant gain only leads to overloading. Those which are available today are generally used to give additional RF tuning. This helps improve aspects of the performance such as overloading from signals well off the received channel and intermodulation. In some units amplification is provided only to overcome the loss caused by the tuning stages.

Radio frequency preamplifiers without any tuning are only used as part of an active aerial. If they are used with full sized aerials they give rise to overloading. This results in a general increase in background noise level and the generation of spurious signals. Usually this reduces the sensitivity of the set, the opposite of the desired effect.

6.3 Audio filters

One way in which the performance of a receiver can be improved is to add additional filters. Sometimes it is possible to fit additional filters into the set itself. However, this is not always possible and it may be decided that the better option is to use an additional filter added on to the outside of the set.

A variety of filters are available. A few offer basic audio filtering, but most filters which are on offer give more comprehensive facilities. One very useful facility is a **notch**. This type of filter allows a small part of the audio spectrum to be removed to eliminate, for example, an offending whistle, as shown in Figure 6.2. Although some of the wanted signal around the offending heterodyne is removed, this does not normally affect the intelligibility. This type of filter is particularly useful when copying SSB, as offending carriers appear as heterodynes which reduce the readability of the signal. Although the notch can be adjusted manually on most filters, some of the more sophisticated versions can detect the offending signal and remove it automatically. These filters are known as **frequency agile notches**.

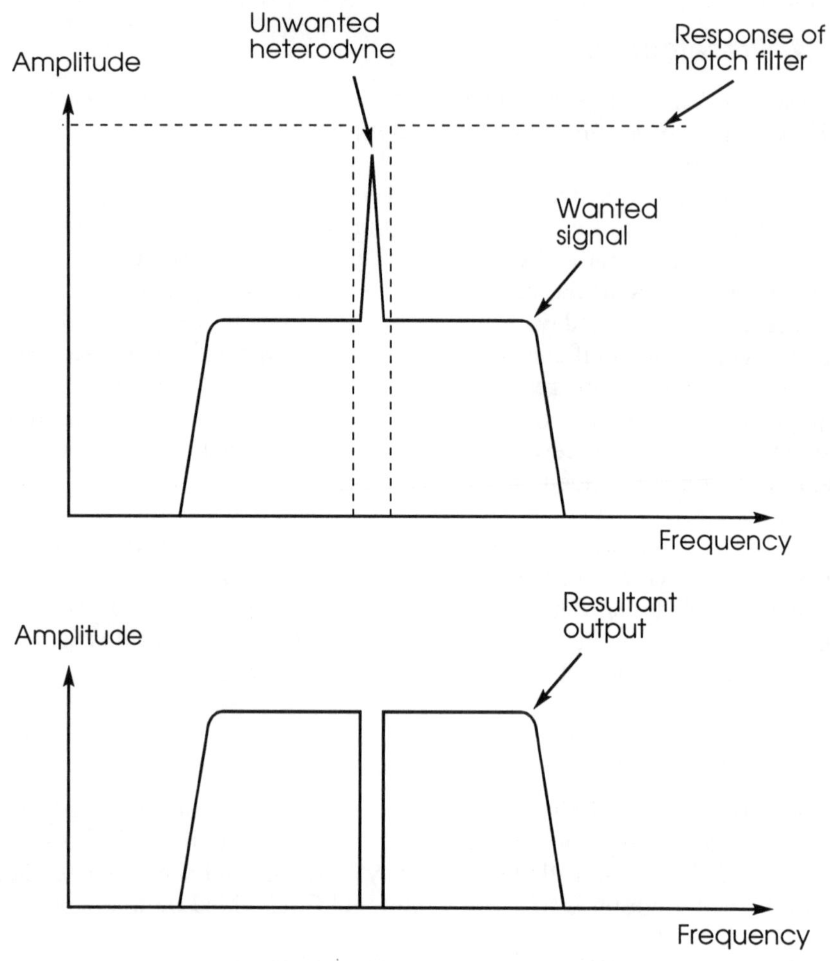

Figure 6.2 *Effect of a notch filter*

A number of new digital signal processing (DSP) filters are appearing on the market. These offer a wide range of facilities and an impressive performance. One of the advantages of a DSP filter is that much greater flexibility is available than with those using conventional circuitry. There are usually a number of fixed positions for high pass, low pass and bandpass filtering. They are also able to offer user selected bandpass filtering where both the centre frequency and bandwidth are selectable. In addition to this, they often offer an automatic frequency agile filter capability.

The main disadvantage of these filters is that, like any audio filter, they are placed after the AGC sensing point. This means that a strong signal which is being rejected by the DSP filter may capture the AGC and reduce the level of the wanted signal. This is particularly important when the level of the signal is varying, as in Morse. Here the level of the wanted signal will vary as a result of the changing level of the AGC making it very difficult to copy. However, these DSP filters can still make a significant contribution to improving copy of many signals.

6.4 Frequency calibrator

Most receivers today are quite accurate. However, there is sometimes a need to check the accuracy of a receiver, particularly older ones which are not so accurate. The easiest way to achieve this is to have an accurate oscillator. If this generates an accurately known low frequency signal, then its harmonics can be picked up throughout the short wave spectrum.

Some calibrators are able to give outputs at 10 MHz, 1 MHz and 100 kHz. These are naturally switched so that only one is on at any given time and each one is made so that the output is rich in harmonics. This means that the 10 MHz signal will give signals at 10, 20, 30 MHz and so forth. Similarly the other outputs will give signals to cover the short wave bands. When checking the calibration of a receiver, first check where the 10 MHz signal appears on the dial, then the 1 MHz signals and last the 100 kHz signals. The checks should be performed in this order in case the dial is a long way out. If the 100 kHz signal was used, first it might be possible to set the receiver so that it was 100 kHz off frequency.

The critical element in a calibrator of this nature is the oscillator. Usually there is an adjustment which can be made on the oscillator and the output can be compared to another known accurate source and set accordingly. Standard frequency transmissions can often be picked up on the short wavebands and these can be used.

Some calibrators pick up low frequency standard frequency transmissions. In Britain MSF at Rugby broadcasts on 60 kHz and can be used. Alternatively the long wave BBC Radio 2 transmission on 198 kHz is maintained to a very level of accuracy for this purpose and it is often used. Naturally calibrators or frequency standards of this nature are more expensive. For many people a calibrator of this standard is not necessary in view of the cost. Most of today's receivers are quite accurate enough, and where an older receiver is in use which may need a calibrator, a crystal controlled one which can be manually set to a known standard is usually quite sufficient.

6.5 Image and data decoders

In order to be able to copy data, facsimile and slow scan television signals specialized units need to be connected to the audio output of the receiver to decode the signals. Data transmissions are very popular with transmitting radio amateurs and there are also plenty of slow scan amateur transmissions as well as the commercial facsimile signals.

RTTY is still widespread. Some years ago it was received by the aid of large teleprinters or teletypes. These were widely available on the surplus equipment market for modest prices, making an RTTY station easy to set up. Although other modes are gaining in popularity, it is still widely used.

Packet, AMTOR, PacTor and SITOR are growing in popularity and their use is widespread. However, when receiving some of these modes there are a number of points to note. Packet is designed to give error free reception by a station which can send acknowledgements. This means that a listener who cannot transmit any acknowledgements and is listening to a link which has already been established may receive a rather garbled message. This generally occurs if parts of the message have to be repeated as the same packet will be received more than once. Similar problems may also be experienced when listening to AMTOR, PacTor and SITOR.

Slow scan television is also very popular. In the early days, cathode ray tubes with long delay times were used to display the image. These were far from ideal as the top of the picture would start to fade before the bottom was complete. Special machines were also used for facsimile.

There are a number of more up to date methods for decoding these signals. The first is to use a simple hardware interface to convert the audio tones from the receiver into voltages which can be processed by a computer. With the widespread use of computers today this offers an ideal solution to many people. By using the interface and suitable software good results can be achieved with the minimum of outlay. A variety of signals from packet and slow scan television to facsimile and AMTOR or SITOR can be received by these decoders. However, some of the interfaces and software packages can only decode a limited number of modes.

A variety of packages are available giving a host of facilities. Some are more biased towards data communications whilst others give a number of facilities which are more likely to be used by the short wave listener who is interested in the reception of weather maps or other transmissions. In particular there are a number of packages which are designed particularly for weather information reception. These can represent good value for money because they are widely used by a variety of people from short wave listeners to owners of sea-going craft.

Software is available for a variety of computers including many of the older types like the BBC, Apple 2 and various varieties of Sinclair computers. Today most of the software is being written for the Apple Macintosh or IBM PC. Much of this software is available as shareware. However, it is worth remembering that once the choice has been made, the full fee should be paid. If this is not done then software writers will not have the incentive to keep up with the latest developments and listeners will lose out in the long term.

Whilst many people will want to opt for an external interface, some may consider a card which plugs directly into the computer. This has a number of advantages as it allows the computer more direct access to the demodulation process, giving a much greater degree of flexibility. Unfortunately it does have the disadvantage that the interface can use up to two slots within the computer. For some people with a number of additional boards already in the computer this may be a problem.

A number of multimode controllers are also available. Most of them cater for a variety of signals from packet to facsimile and even Morse. Again, a computer may be used to display the signal being received, but often it may be sent directly to a normal television or a printer for viewing without the need for additional hardware.

Many controllers use traditional techniques, but a number which use digital signal processing are coming onto the market. This new approach can be put to good use to make even more flexible units. A few of them are capable of acting as filters as well as decoders. Here the advantage of digital signal processing has been exploited, enabling a whole variety of facilities to be incorporated by the addition of additional software. Although many of these units are still fairly expensive, they are likely to become more widely used in the future, especially as they offer extremely good performance.

Whatever option is considered, there is a wide choice of equipment available for this growing section of the market. It is worth taking some time to look at all the options which are available and choosing what will be best for each particular station.

6.6 Clocks

A good reliable clock is a very useful item in any short wave listening station. Often an accurate time check is needed. For example, when listening to broadcast stations it is necessary to know the time relatively accurately because station names and interval signals are generally broadcast close to the hour or half hour or sometimes quarter past and quarter to the hour.

A quartz clock will keep good time and can be checked using the short wave broadcasts to ensure it is always correct. An alternative is to buy a clock which is locked to one of the low frequency standard frequency transmissions. These can be picked up in most areas of Western Europe and North America, although it is best to check about the reception before buying one of these clocks.

The transmissions give out additional data which enable these clocks to correct themselves. Usually there is a facility which allows the time to be changed as the clocks change between summer and winter times. However it may be more useful to keep it set to GMT (UTC), especially in Western Europe, as it is the standard used by several broadcasters and all time zones are measured relative to it. Having an easy indication of GMT can also help when logging stations and sending out reports. Using GMT enables a common standard to be adopted regardless of the time zone of both stations.

7 Users of the radio spectrum

The radio spectrum is required for a vast number of different uses. Broadcasters, ships, aircraft, military and a host of other organizations all need to be able to use radio. Each has its own requirements and often they are not compatible with one another. Different modes of transmission may be needed and some will need to use very high power whilst others will only need low power. To enable all the users to be able to use the radio spectrum with the minimum of interference, different bands of frequencies are allocated to different users. Some bands are allocated for one purpose only, whereas other bands are shared by two or more users.

The radio spectrum covers an extremely wide band of frequencies. At the low end of the spectrum frequencies of a few tens of kilohertz are used for navigational purposes. At the high end of the spectrum scientists are pushing forwards the bounds of technology so that semiconductors can operate at frequencies of 100 GHz (100 000 000 000 Hz). Between these two extremes all the commonly used forms of transmission can be found.

The short wave bands are generally considered to extend from frequencies around 1.5 or 2 MHz up to 30 MHz. As such it extends from the MF portion of the spectrum to the top of the HF section. Above this are the VHF and UHF areas used for VHF FM radio broadcasts and television transmissions.

With the large number of people wanting to use the radio spectrum, different areas are allocated to different types of user. In this way the best possible use is made of the available spectrum. These allocations are made by international agreement. Periodically World Radio-communication Conferences (previously called World Administrative

Radio Conference or WARC) are held to discuss aspects of the allocations to meet the changing requirements.

The administration and running of these allocations and other related matters is undertaken by the International Telecommunications Union (ITU). It allocates callsigns series to countries for circumstances where radio stations need callsigns. For managing the radio spectrum it has also split the world into three regions. This has been done because the requirements vary from one area to another. Region one covers Europe, Russia and Africa. Region two covers North, Central and South America, and finally region three covers Asia and Oceania.

7.1 Listening and the law

Whilst there are many types of station which can be picked up, it is not necessarily legal to listen to them. In the UK the law relating to this is contained in the Wireless Telegraphy Act of 1949, section 5 (b). This allows the general public to listen to transmissions intended for general reception. This includes broadcast stations, amateur radio and Citizen's Band signals as well as standard frequency transmissions. Other signals in this category include weather faxes, weather information for mariners and a few other transmissions. Whilst some may be of little interest, others are fascinating to receive.

Two way transmissions intended only for reception by a particular receiving station cannot legally be received in the UK. This means that messages from ships to shore stations, aircraft and the like should not be received.

It is important to adhere to the law as people have been prosecuted, even in the UK. In some countries similar laws are applied far more strictly. In the USA where there is a greater degree of freedom of information, laws relating to reception of unauthorized signals are applied more rigidly. In fact wherever one is located it is necessary to find out what the laws are and adhere to them.

7.2 Summary of users

There are many types of transmission which can be heard on the short wave bands. Radio amateurs and broadcasters are the two traditional areas for short wave listeners. These are covered in the following two chapters. However, there are a number of other users and it is of interest to the short wave listener to know who they are and what the bands are used for.

7.2.1 Maritime

One of the major users of the HF radio spectrum is the maritime service. Its transmissions cannot normally be received legally as they do not make transmissions for general reception. Only weather broadcasts are an exception to this rule.

The service uses the HF bands chiefly to maintain communications between the ship and shore. Radio is the only viable means of communication between the ship and the outside world. Whilst VHF communications is used for much local work, the HF bands are used for much of the longer distance traffic. Even though the use of satellites is growing for this type of communication, the short wave bands are still widely used and will continue to be for many years. Communications using these bands is much cheaper for two reasons. Firstly, there is no charge for the ionosphere, whereas satellites are expensive and charges are made by the operators. Secondly, the equipment for HF communications is readily available and can often use much older technology which again makes it cheaper. Obviously many new and prestigious vessels use satellites, but there are many older and smaller craft on the high seas which need to keep costs to a minimum.

There are several types of transmission which appear on the bands. Often coastal stations can be heard giving what are called traffic lists. These consist of general information to all stations. The weather information has already been mentioned as it can be legally received. Other information also includes general navigational information and warnings.

Coastal stations are also used for what are called link calls where a ship needs to make a ship to shore telephone call. Here the shore station links the ship's call through to the telephone network so that calls can be made to a variety of people. One example of a call may be when the ship's captain needs to talk to the owners to inform them of estimated times of arrival in a port, or other essential information.

Sometimes ships need to talk to one another. This is particularly true of fishing ships which often talk to one another comparing catches and talking about other general information. This is more common on the MF bands as the distances which need to be covered are much smaller.

Finally, distress calls are occasionally made. A number of calling or distress frequencies are reserved. These are normally used for calling shore stations. Once contact has been made, the stations move off to another frequency so that it remains clear. Calling channels are also used for distress signals. The most active calling channel is 2182 kHz and a radio silence is kept on this channel for three minutes after the hour and half hour. The radio silence is kept because the channel is normally very busy and keeping it clear at regular intervals enables weak signals to be detected more easily. This is particularly important because many life craft carry small battery powered distress transmitters which will only give out a weak signal.

7.2.2 Aeronautical

Aircraft make widespread use of the HF bands. Like the maritime service they cannot be received legally in the UK.

The amount of long distance traffic between aircraft and the ground has steadily grown over the years. Not only has the number of aircraft grown dramatically in the past few years, but with the airways becoming more congested, the need to maintain communications with aircraft has also increased.

Much of the communication between an aircraft and the ground takes place in the VHF aeronautical band. Because of the frequency involved relatively short ranges are possible. This is fine when the aircraft is over a populated area like Europe and Northern America. Here the aircraft will be in VHF range of an airport or control for most of the time and longer range communications will not be needed. However, when crossing areas like the Atlantic or sparsely populated areas of the world the aircraft will be outside the range of VHF stations for long periods of time. Whilst satellite communication is increasing, there is still heavy reliance on the HF bands.

One of the primary needs to maintain contact with the ground is so that the position of all aircraft can be monitored and corrected if necessary. Accordingly there are a number of stations placed at either side of the Atlantic for this purpose. These busy stations control the traffic crossing the Atlantic on what are called the Atlantic tracks. Aircraft regularly report their latitude and longitude together with other essential details including their speed, altitude and information about the wind speeds they are encountering.

Often the crew on an aircraft may need to talk to their own company personnel. There are many instances when this may happen. For example, they may need to organize maintenance to the aircraft ahead of landing or to report a new estimated time of arrival. Messages like these can be sent in a number of ways. Some of the larger airlines own their own stations and the aircraft can talk directly to them. Alternatively a ground station can be used to forward messages or even connect them through to the telephone network when they can talk directly to the ground staff.

7.2.3 Volmet stations

These stations are also called Meteos and transmit weather information, primarily for aircraft. They enable pilots to gain a good idea of the weather at their destination. Some stations transmit for brief periods of time whereas those at important points normally transmit continuously. In some instances two stations may share the same frequency. When this happens they normally time share so that only one station is transmitting at any given time. The information carried is normally figures about the weather conditions and this is repeated every few minutes.

7.2.4 Meteorological maps

There is a vast amount of information transmitted each day by meteorological agencies. Many of these transmissions can be received quite legally in the UK because these transmissions are intended for general reception. Much of the output from the UK is transmitted by the Meteorological Office which is located in Bracknell in Berkshire, to the west of London.

Information is transmitted in a variety of forms. A number of transmissions are also made in RTTY. Although more up to date systems like facsimile are available, RTTY systems are still in use in a number of installations and as a result these transmissions are still made. However the number of RTTY transmissions is steadily decreasing in favour of the more up to date standards which give better information and more informative images.

Facsimile is one of the most interesting forms of weather information transmissions. It is easy to interpret with the correct equipment, and the images which can be received are fascinating. Bracknell regularly transmits maps of Europe and the North Atlantic, as well as maps showing ice floes for even the western side of the Atlantic. In winter these show areas as far north as Iceland and other areas in the vicinity. Many of these maps show the cloud formations, whilst some show the high and low pressure areas and others give general meteorological information.

A variety of drum speeds are used. The most common is 120/576, although other standards are used. Russian stations tend to favour 060/576, 090/576, 060/288 and 090/288 formats. These transmissions are found throughout the short wave spectrum, so that despite the propagation conditions it will be possible for receiving stations like ships to be able to receive at least some signals to give them warning of the weather conditions. To enable stations to know when transmissions are to be made and on which frequencies, many stations transmit schedules either for the day ahead and some even give schedules as far ahead as a month. Once these have been received it gives a good guide about future transmissions. Although many frequencies change, Bracknell can usually be heard on 2.6185, 4.61, 8.04 and 14.436 MHz. Other stations also have regular frequencies on which they can be heard. The up to date frequencies can be obtained from one of the frequency listings which are published.

Some transmitters like those in Bracknell operate 24 hours a day. Even though they may not transmit information all the time, a carrier is present continually. Other stations are not on the air all the time and it is more difficult to know whether propagation conditions will allow reception.

7.2.5 Press agency information

A number of countries transmit press releases which are intended for general reception. However, any press releases transmitted for a particular station or group of stations should not be received; these are often sent in an encrypted form. Most news services use transmissions at 50 baud having a 425 Hz frequency shift whereas some use 57 baud with the same frequency shift. Most of the transmissions use RTTY as this can be copied by a wider audience, whereas some are SITOR B to give more reliable and error free copy.

Signals can be heard from a variety of sources all over the world. Often the stations give a schedule of their transmissions enabling further bulletins to be copied at a later time or date. These schedules are particularly useful because the frequencies and times of transmission change periodically to enable the best to be made of the propagation conditions. However most transmissions are on frequencies between about 4 and 20 MHz. These normally give the most reliable communication and reach a sufficiently wide area. The frequencies can be obtained from one of the frequency listings which are published.

7.2.6 Citizen's Band

Citizen's Band (CB) is allowed in a very large number of countries around the world. Although normally identified with the USA where it was started, it is now firmly established in all parts of the world. These transmissions can be received by the short wave listener in the UK as they are classed as being intended for general reception.

Unlike the amateur licence which requires tests to be passed before the licence is issued, the CB licence is open to all. There are restrictions. In the UK equipment has to be approved and cannot be modified. Transmitter power is also limited. Limits of up to 5 watts are generally imposed, although the actual limits vary slightly from one country to the next.

CB is used by a wide variety of people. Some just want to talk over the radio, keeping in touch with friends. Other people use it as a very convenient means of short range communications. Many lorry drivers use it as a companion in the cab as well as a means of finding out about road conditions.

Soon after the introduction of CB in the USA many people started to use it for making long distance contacts. Although it is illegal to make contacts over a distance of 150 miles in the USA, this rule was difficult to enforce and many people disregarded it. Even now many people still make long distance contacts, sometimes with people outside the country. For the listener it is often possible to hear stations over considerable distances. Sometimes it is possible to hear stations over several thousand miles at the peak of the sunspot cycle, despite the low power levels.

Although the conditions for CB operation vary from one country to the next, many countries are falling more into line with one another. The channels used across Europe fall into line with those used in the USA. In the UK, the original channels are now being phased out and the new European channels are being used increasingly. As the occupancy of channels increases this does unfortunately make hearing long distance stations more difficult.

7.2.7 Standard frequency signals

There are a number of standard frequency transmissions which are broadcast from around the world. The frequency of these transmissions is maintained very accurately, but their main use for the short wave listener is as a convenient indicator of propagation conditions. A number of stations can be heard around the world, each giving its callsign at regular intervals so that it can be identified. These callsigns fall in line with the internationally agreed set of prefixes given in the appendix and they usually identify themselves in Morse.

One of the most famous HF transmissions comes from WWV located at Fort Collins, Colorado. It has a sister station with the callsign WWVH located in Hawaii. These stations can be heard giving the time every minute and propagation information at 15 minutes past the hour in speech.

Signals are not always radiated continually because several stations have to share the same frequencies. As identification these transmissions use Morse to give their callsigns, usually at the beginning of each period of transmission. Thereafter the time pips are usually transmitted.

The Russian station RWM can be heard very easily in the UK. It transmits just below the round frequencies to alleviate interference. It is 4 kHz below 5.00, 10.00 and 15.00 MHz. Similarly the station RID in Irkutsk is 4 kHz above these frequencies making reception easier.

There are also a number of standard frequency transmissions which are much lower in frequency. MSF is one such station, transmitting on 60 kHz from Rugby in the UK with a power of 50 kW. The frequency of this station is traceable back to the standards at the National Physical Laboratory in Teddington, England, and the frequency is maintained to an accuracy of better than ±2 parts in 10^{12}. A similar station with the callsign DCF transmits a signal on 77.5 kHz. These low frequency transmissions are often used to maintain the precision of high accuracy oscillators used as laboratory standards.

All these stations are maintained to a very high degree of accuracy and are more than accurate enough for the average short wave listener. However, for the professional user the phase changes of the received signal caused by a number of effects in the transmission path may mean that the short term accuracy is not sufficient for all the requirements. Nowadays more people are using the standards received from GPS (Global Positioning System) as they have a higher level of accuracy.

7.2.8 Other users

Apart from the users which have already been mentioned, there are many other organizations which need access to the radio spectrum. Naturally the military of all nations need efficient radio systems to keep in contact with one another and they make good use of the spectrum. Diplomatic services also need radio systems. For example, embassies in countries often have large HF aerial arrays on the roof. Naturally the military and diplomatic services encrypt most of their messages so that their content cannot be copied.

A number of experimental licences are also issued. These are used for a variety of purposes from developing new radio equipment to making transmissions to discover more about the ionosphere.

There are many other uses for the HF spectrum. Unfortunately the listener should not listen to many of them. However there is more than enough which can be heard legally to keep the listener fascinated and enthralled for many hours each week.

Table 7.1 Original UK 27 MHz Band Channel Frequencies (MPT1320 (27/81))

Channel	*Frequency*	*Channel*	*Frequency*
1	27.60125	21	27.80125
2	27.61125	22	27.81125
3	27.62125	23	27.82125
4	27.63125	24	27.83125
5	27.64125	25	27.84125
6	27.65125	26	27.85125
7	27.66125	27	27.86125
8	27.67125	28	27.87125
9	27.68125	29	27.88125
10	27.69125	30	27.89125
11	27.70125	31	27.90125
12	27.71125	32	27.91125
13	27.72125	33	27.92125
14	27.73125	34	27.93125
15	27.74125	35	27.94125
16	27.75125	36	27.95125
17	27.76125	37	27.96125
18	27.77125	38	27.97125
19	27.78125	39	27.98125
20	27.79125	40	27.99125

Table 7.2 CEPT and USA 27 MHz Band Channel Frequencies (MPT 1333 (PR 27/GB))

Channel	*Frequency*	*Channel*	*Frequency*
1	26.965	21	27.215
2	26.975	22	27.225
3	26.985	23	27.255
4	27.005	24	27.235
5	27.015	25	27.245
6	27.025	26	27.265
7	27.035	27	27.275
8	27.055	28	27.285
9	27.065	29	27.295
10	27.075	30	27.305
11	27.085	31	27.315
12	27.105	32	27.325
13	27.115	33	27.335
14	27.125	34	27.345
15	27.135	35	27.355
16	27.155	36	27.365
17	27.165	37	27.375
18	27.175	38	27.385
19	27.185	39	27.395
20	27.205	40	27.405

8 Amateur radio

There is a very large band of people interested in radio who also like to transmit. These radio amateurs or 'radio hams' are able to communicate with fellow enthusiasts all over the world, conversing freely about all manner of topics. A thoroughly fascinating development of short wave listening, many thousands of people hold licenses today and the level activity of is growing steadily.

There are many aspects to amateur radio. Listening on the bands it is possible to hear people talking about their interests in the hobby. Some enjoy chatting to folk around the globe and making friends in different countries or different continents. Others enjoy the challenge of chasing stations in new countries, increasing the total of countries with which they have made contact. Many people enjoy the technical side of the hobby, making their own equipment or experimenting with aerials to improve their signal. With all of these interests there is a wide variety of people to listen to, and it is interesting getting a flavour of the different countries around the world from what the people say. However, in just the same way that radio amateurs have their own brand of the hobby, the same is also true for listeners. It is possible to make the hobby what one wishes, and change the flavour from time to time. In this way the hobby can become a life long pastime with always something new to ensure that it is interesting and stimulating.

8.1 Licences

Whilst many people are content just to listen to radio amateurs, some will want to transmit. To be able to do this it is necessary to hold a licence. These are obtained from the relevant government agency. In the UK, the Radio Regulatory Department of the Department of Trade and Industry issues amateur licences. Once a licence has been obtained it is possible to transmit and take part in many of the amateur activities in a new way.

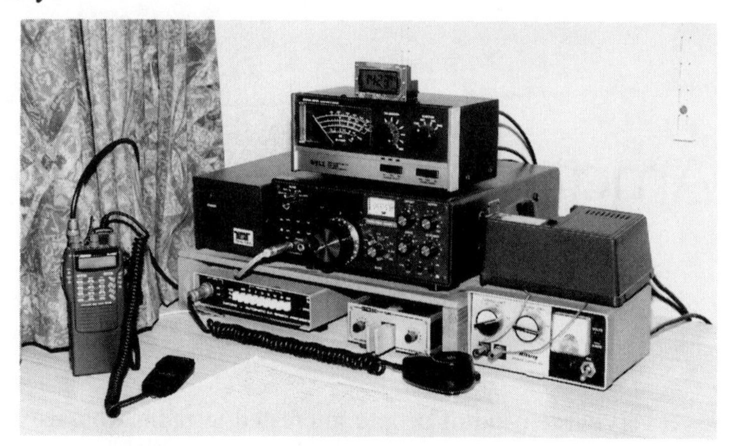

Figure 8.1 *A typical amateur radio station*

The requirements for obtaining licences vary from one country to the next. In general it is necessary to pass a theory examination. It is then possible to obtain a licence for operation above 30 MHz. To operate on the short wave bands below 30 MHz it is necessary to pass a Morse test as well.

In the UK, the theory examination is called the Radio Amateurs' Examination and it is run by the City and Guilds of London Institute. It covers a broad base of basic radio theory and the licence conditions. These ensure that people allowed onto the amateur bands have a basic knowledge of the essential topics. The Morse test, which enables people to apply for a licence for operation on the short wave bands as well, is administered by the Radio Society of Great Britain and consists of sending and receiving at 12 words a minute.

In addition to the standard licences, novice licences are also available. The requirements for obtaining them are less than for the normal licence, although transmitter powers and frequencies are more limited. Aimed at the younger enthusiast, they give an ideal introduction to the hobby and they are a good stepping stone for progressing on to a full licence. Whilst novice licences are aimed at younger enthusiasts, there is no reason why people of any age cannot take one out.

Further details of all the amateur licences in the UK can be obtained from the Radio Society of Great Britain, Lambda House, Cranborne Road, Potters Bar, Hertfordshire EN6 3JE. Similar licences are available in most countries and usually there is a national society of relevant government department which can furnish all the details of this exciting hobby.

8.2 Callsigns

When a station is issued with a licence, it is given its own unique callsign. This enables the station to be identified when it is operating on the air. It is normally used at the beginning and end of each transmission.

Callsigns can be split into two sections: the prefix and the serial letters. The prefix consists of the characters up to and including the last number. The serial numbers are the remaining letters of the callsign and they act as an identifier for that particular station within the prefix group.

Each country is allocated prefixes which it can use and by reference to a list it is possible to identify the country and sometimes a district within the country where the station is located. A list of the prefixes used by amateur stations is given in the appendix.

An example of a typical callsign is G3YWX. In this the prefix is G3 and from the list of amateur prefixes it can be seen that the station is located in England. Similarly the station VP8ANT is either in the Falkland Islands, South Georgia, South Shetland Islands, South Sandwich Islands, South Orkney Islands or the Antarctic. In fact this particular station was located in the Antarctic.

During contests or when stations are on the air for special events, stations may use different prefixes so that they can attract more attention. Whilst these prefixes may not be within the normal amateur lists, they still fall within the internationally agreed prefixes allocated to countries for all types of stations. These prefix allocations are also given in the appendix.

Often suffixes are added to a callsign. When a station is operating from an automobile it will use the suffix /M for mobile. The suffix /P may be used for portable operation, and /MM for maritime mobile. Occasionally /AM will be heard when a station is aeronautical mobile.

With international travel more common these days, radio amateurs often want to set up stations in other countries. If they anticipate being there for some while a totally new callsign for the country in question is usually allocated. However if a short stay is envisaged the station is often not given a new callsign. Instead the home callsign is used with the prefix of the country where the station is located. Traditionally this additional prefix was added after the main callsign. For example G3YWX/VP9 would be located in Bermuda. However, there is now a trend to place the additional prefix before the callsign. F/G3YWX would be located in France.

8.3 Jargon and codes

To the newcomer to amateur radio there may appear to be a large amount of jargon used in normal communications. This has come about for a number of reasons. In the first instance there is the normal technical jargon associated with radio and electronics.

Table 8.1 Common Abbreviations

ABT	About	LID	Poor Operator
AGN	Again	MOD	Modulation
AM	Amplitude Modulation	ND	Nothing Doing
ANT	Antenna	NW	Now
BCI	Broadcast Receiver Interference	OB	Old Boy
BCNU	Be Seeing You	OM	Old Man
BFO	Beat Frequency Oscillator	OP	Operator
BK	Break	OT	Old Timer
B4	Before	PA	Power Amplifier
CFM	Confirm	PSE	Old Timer
CLD	Called	R	Roger (OK)
CONDX	Conditions	RCVD	Received
CPI	Copy	RTTY	Radio Teletype
CQ	A general call indicating a contact is wanted	RX	Receiver
		SA	Say
CU	See you	SIGS	Signals
CUAGN	See You Again	SRI	Sorry
CUD	Could	SSB	Single Sideband
CW	Continuous wave (used to denote a Morse signal)	STN	Station
		SWL	Short Wave Listener
DE	From	TKS	Thanks
DX	Long Distance	TNX	Thanks
ERE	Here	TU	Thank You
ES	And	TVI	Television Interference
FB	Fine Business	TX	Transmitter
FER	For	U	You
FM	Frequency Modulation	UR	Your, You Are
FONE	Telephony	VY	Very
GA	Good Afternoon	WID	With
GB	Goodbye	WKD	Worked
GD	Good	WUD	Would
GE	Good Evening	WX	Weather
GM	Good Morning	XCVR	Transceiver
GN	Good Night	XMTR	Transmitter
GND	Ground	XTAL	Crystal
HBREW	Homebrew (home-made)	XYL	Wife
HI	Laughter	YL	Young Lady
HPE	Hope	Z	GMT — added after the time, e.g. 1600Z is 16.00 GMT
HR	Here		
HV	Have	73	Best Regards
HW	How	88	Love and Kisses

Apart from the technical talk, there is a wide selection of abbreviations and codes which form part of the amateur radio vocabulary. These have developed over the years. Many were introduced primarily to help sending messages quickly and efficiently. Although they are still widely used for Morse transmission, they have also been incorporated into speech as well, even though they do not provide any real time saving.

Many abbreviations are quite obvious. Others may just change a letter in the word. This is generally to make it easier to send in Morse. For example the word 'for' is changed to 'fer'. This is because the letter 'O' consists of three dashes in Morse and is much longer to send than the letter 'E' which is a single dot. Others like AM and FM are standard abbreviations which are often seen elsewhere.

Another set of codes which is widely used is called the Q code. Again this was initially introduced to help in Morse transmissions to handle long but frequently used questions and answers quickly and concisely. Each code consists of a three letter string beginning with the letter Q and can be used as either a question or an answer. If a query is placed after the code then this indicates it is being used as a question, whereas the code on its own or in some cases with further information indicates a reply.

Table 8.2 The Q Code

Code	Meaning
QRA	What is the name of your station? The name of my station is ....
QRG	What is my exact frequency? My exact frequency is ....
QRK	What is the readability of my signal? The readability of your signal is ....
QRL	Are you busy? I am busy.
QRM	Is there any (man-made) interference? There is (man-made) interference.
QRN	Is there any atmospheric noise? There is atmospheric noise.
QRO	Shall I increase power? Increase power.
QRP	Shall I decrease power? Decrease power.
QRQ	Shall I send faster? Send faster.
QRS	Shall I send more slowly? Send more slowly.
QRT	Shall I stop sending? Stop sending.
QRU	Do you have any messages for me? I have no messages for you.
QRV	Are you ready to receive? I am ready to receive.
QRZ	Who is calling me? You are being called by ....
QSK	Can you hear between your signals? i.e. use break in on Morse transmissions. I can hear between my signals.
QSL	Can you acknowledge receipt? I can acknowledge receipt.
QSP	Can you relay a message? I can relay a message.
QSY	Shall I change to another frequency? Change to another frequency.
QTH	What is your location? My location is ....

The codes have very specific meanings which are included in publications including the handbook issued by the International Telecommunications Union (ITU). Many of the codes are used by specific services including maritime and aeronautical users. Those listed in Table 8.2 are widely used by radio amateurs.

In view of the widespread use of the Q code it not always used in its strict question and answer format, especially when it is used for speech transmissions. Here the codes are often incorporated into the normal vocabulary. For example it is quite common to talk about a QRP transmitter when meaning a low power transmitter or to mention that there is QRN on the band meaning there is static interference. However, when these expressions are used, their basic meaning is retained.

It is often difficult to distinguish between the sounds of different letters over the air. Often it is necessary to give letters when spelling out a name or giving a callsign. Letters like 'B' and 'P', or 'F' and 'S' can easily be mistaken even when conditions are good. With interference and the reduced audio bandwidth of the transmitter and receiver, mistakes are very easily made. To overcome this a phonetic alphabet is used. The one shown in Table 8.3 has been adopted by the International Telecommunications Union and is internationally agreed. Using this, the word Leeds would be spelt out as Lima Echo Echo Delta Sierra, thereby avoiding the possibility of any confusion.

Even though the alphabet given in Table 8.3 is internationally accepted, many radio amateurs still use their own versions. A common variation is to use famous cities like Yokohama, Washington and the like. However X-ray tends to remain the same whatever system is used!

Table 8.3 International Phonetic Alphabet

A	Alpha	N	November
B	Bravo	O	Oscar
C	Charlie	P	Papa
D	Delta	Q	Quebec
E	Echo	R	Romeo
F	Foxtrot	S	Sierra
G	Golf	T	Tango
H	Hotel	U	Uniform
I	India	V	Victor
J	Juliet	W	Whisky
K	Kilo	X	X-ray
L	Lima	Y	Yankee
M	Mike	Z	Zulu

8.4 Signal reports

When radio amateurs are in contact, it is very useful to report how the other station's signal is being received. These reports enable both stations to tailor the contact according to the conditions. If readability is poor and signal strengths are low, then the contact can be kept short so that communication is not lost. Reports also give a guide to such factors as the band conditions and how well the station is operating. If a transmitting station consistently receives poor reports it may mean there is a problem with the equipment.

To give meaningful indications about the signal it is necessary to have a concise method of giving the report. The system which is universally used by radio amateurs is given in Table 8.4. Called the RST system it gives indications about readability, strength and tone. There are five levels of readability, whereas the strength and tone have nine.

For speech contacts only the readability and strength indications are used. For Morse the tone is included as well. This was very useful some years back when more equipment was home constructed. Today most signals are of a high standard, although there are occasionally some which have problems.

Typically a report on a speech signal may be five and eight for a totally readable strong signal. A Morse signal may be 579 if it is totally readable, moderately strong with a pure DC note. Sometimes the reporting of numbers in Morse may be shortened. Nine being four dashes and a dot can be shortened to a dash and a dot or the letter N. In this case a report of 599 may be shortened to 5NN.

Table 8.4 RST Code for Readability, Strength and Tone

	Readability		*Strength*		*Tone*
R1	Unreadable	S1	Barely detectable	T1	Extremely rough note
2	Barely readable	2	Very weak signals	2	Very rough note
3	Readable with difficulty	3	Weak signals	3	Rough note
4	Readable with little difficulty	4	Fair signals	4	Fairly rough note
5	Perfectly readable	5	Fairly good signals	5	Note modulated with strong ripple
		6	Good signals	6	Modulated note
		7	Moderately strong signals	7	Near DC note but with smooth ripple
		8	Strong signals	8	Near DC note but with trace of ripple
		9	Very strong signals	9	Pure DC note

It is not always easy to give an accurate strength report. The human ear is very insensitive to changes in the level of a sound. The smallest change which the ear can usually detect is about 2 dB (i.e. when the signal level has changed by a factor of about 1.5:1), and this makes giving accurate strength levels very difficult. To help overcome this most short wave receivers have strength meters, which are usually calibrated in 'S' units.

For an S meter reading to be meaningful it must relate to a given change in the signal level. Usually a change of 6 dB is considered to be equivalent to one S point, although in some cases 4 dB is used.

Even though the changes for S points are defined, the meters themselves are usually very inaccurate because of the nature of the circuitry in the set. This means that the inaccuracy of these meters must be borne in mind when using them. If they are used as a comparison between different signals they can be very useful.

8.5 Amateur bands

Radio amateurs have access to a number of bands of frequencies within the HF portion of the spectrum. Some of these bands, particularly those higher in frequency, are allocated only to radio amateurs, whereas some of the lower frequency bands are shared with other services.

A number of the bands used today by radio amateurs have been allocated to amateur radio since before 1945. However in recent years a number of new bands have been released for amateur use. Many of these changes were agreed in the World Administrative Conference held in 1979, although it has taken some time for the previous users to move to their new allocations, relinquishing the new amateur bands. As a result these bands have only been released for amateur use in the past few years.

Although the allocations are broadly the same worldwide there are some differences between the three ITU regions and different countries. The UK allocations are given in Table 8.5.

Novices in the UK are allowed to operate on a limited number of frequencies in the short wave bands as shown in Table 8.6. Powers are also limited. A maximum output power of 3 watts or input to the final device activating the aerial of 5 watts is allowed. This is comparatively small when compared to the maximum level of 400 watts which is allowed for holders of full licences. However, these levels of power are quite sufficient to make contacts all over the world.

Table 8.5 UK Amateur Bands		
Frequency Limits (MHz)		*Approximate Wavelength and Band Name*
1.81	2.0	160 metres (Top Band)
3.50	3.80	80 metres
7.00	7.10	40 metres
10.10	10.15	30 metres
14.00	14.35	20 metres
18.068	18.168	17 metres
21.00	21.45	15 metres
24.89	24.99	12 metres
28.00	29.70	10 metres

Table 8.6 UK Novice Radio Amateur Allocations	
Frequency Band (MHz)	*Types of Transmission Permitted*
1.950 – 2.000	Morse, Telephony, RTTY, Data
3.565 – 3.585	Morse
10.130 – 10.140	Morse
21.100 – 21.149	Morse
28.100 – 28.190	Morse, RTTY, Data
28.300 – 28.500	Morse, Telephony

The amateur bands are well spread over the HF spectrum, giving a wide choice of propagation and operation. Each band has its own character and the experienced short wave listener will know which band to choose to have the best chance of being able to receive stations from a particular part of the world at any given time of the day.

8.5.1 160 metres (Top Band)

This band is often used for local communication. Being just above the medium wave broadcast band, it has many of the same characteristics. During the day short distances are generally all that is possible. Typically distances of up to 40 or 50 km can be easily achieved dependent upon the aerials and powers being used.

At night time when the D layer disappears much greater distances are possible. Often stations up to distances of 500 to 1000 km can be heard.

Sometimes transatlantic stations can be heard and on occasions even stations from the other side of the globe may be audible.

One of the disadvantages of this band is that large aerials are usually needed, particularly for transmitting. As a quarter wavelength aerial is about 130 feet or 40 metres, many people do not use the band because full size aerials are too large, especially where only a small garden is available.

Static is another problem. Noise levels can be relatively high and this can make operation less enjoyable, especially if it is only needed for a local contact.

These problems mean that even though top band used to be very popular for local contacts, the level of activity is now much lower, although some stations can be heard on most evenings.

8.5.2 80 metres

This band is widely used for medium distance contacts. During the day stations up to distances of 200 or 300 km can be heard. It is often used as a longer distance 'natter' band.

During the hours of darkness stations from much further afield can be heard. It is not uncommon to hear transatlantic stations, sometimes at very good strengths. Particularly in the spring and autumn at dawn and dusk, stations from the other side of the globe can be heard, often at strengths more usually expected on the higher frequency bands. However, these conditions may be relatively short lived, lasting only half an hour or so.

Conditions on the band are often difficult. In view of the fact that it is shared by other services, levels of interference can be very high. Even so it can be very rewarding when distant stations are picked up.

Aerials for this band are less of a problem than for top band. Many trapped verticals designed to cover the amateur bands cover 80 metres, although often with only a narrow bandwidth. Even so they can perform quite well. Even quarter wavelength end fed wires or dipoles can be accommodated more easily as they are only half the size of those for 160 metres. This means that levels of activity are much higher and a number of stations can always be heard.

8.5.3 40 metres

This band can produce some very interesting signals. Although relatively small, only 100 kHz wide in regions one and two, and levels of interference can be high, it is often under estimated as a long distance band.

During the day stations up to distances of 1000 or 2000 km can usually be heard. As night approaches these distances increase and after dark it is possible to hear stations from all over the globe at different times.

Naturally the seasons affect the band. In winter it is often possible to hear long distance stations during the day, whilst in summer it is usually necessary to wait for nightfall.

In regions one and two the 40 metre broadcast band occupies the frequencies between 7.100 and 7.300 MHz, but some broadcast stations illegally use frequencies just outside the band and can be heard in the amateur band below 7.1 MHz. This practice is much less widespread than it used to be, although it still occurs.

8.5.4 30 metres

This band was one of those released for use by radio amateurs after the World Administrative Conference in 1979. It is still shared with other services, and only used for Morse transmissions. Unfortunately activity is still relatively low, although there are many stations of interest which can be heard. It is not available for use by all countries.

Another reason for the low level of activity is that many of the commercially made aerials for radio amateurs do not cover this band. Trapped verticals and the like tend to cover the more traditional bands of 10, 15, 20, 40 and possibly 80 metres. However some of the new aerials being launched onto the market do have the capability of covering the new bands. As more traps or other features are required to cover the additional bands, they are usually more expensive.

8.5.5 20 metres

This is undoubtedly the major long haul band for radio amateurs. Stations can be regularly heard from places all over the world. Although it is affected by the time of day, the seasons and the sunspot cycle, it is usually possible to hear some stations of interest.

During the day most of the stations tend to be within distances of about 3000 km, although stations which are further afield may be heard quite often. This is particularly true in the autumn and spring, or at the peak of the sunspot cycle.

As dusk approaches, the character of the band changes. Stations from the west start to appear in plenty. As night falls, the band may remain open or in winter and at the sunspot minimum signals will fall in strength as the band closes for the night. When this happens only ground wave signals will be heard.

At dawn signals from far afield are again heard, often from the other side of the globe. As the morning progresses daytime conditions return.

In view of its popularity, interference levels can be very high when the band is open to many parts of the world. However, it does mean that there is the possibility of hearing many interesting stations.

To give an indication of conditions on the band, there is a network of beacons which can be found on 14.100 MHz. These time-share the same frequency so that it is easy to gain an idea of where the band is open for propagation. Details of the beacons are included in Chapter 3.

8.5.6 17 metres

This is another of the new bands released after the 1979 Conference. As a result of this and its narrow bandwidth (only 100 kHz) the levels of activity are still relatively low. It is a very interesting band, capable of producing some interesting stations from all over the world.

Its nature is very much a half way house between the more popular and larger 20 and 15 metre bands either side of it. Because it is higher in frequency than the 20 metre band, it is more affected by the level of sunspots, remaining open for shorter periods of time in the winter, and when the sunspot cycle is at its minimum.

Despite its similarity to the bands either side, it is still well worth investigating when 15 metres is starting to close in the evenings, as 17 metres will remain open for slightly longer. During these periods is likely to reveal more long distance stations.

8.5.7 15 metres

This is an excellent band for long distance listening and it is possibly the second most used of the long distance bands. Although it is more severely affected by the sunspot cycle than the 20 metre band, it is capable of providing many hours of interesting listening.

The band is not open as early in the day as 20 metres. When the band does start to open, stations from the east are generally heard first of all. As the day progresses stations from the north or south start to appear together with some from the west. As the western stations appear, it is generally found that the longer distance ones from the east fade. As night approaches the stations from the west dominate, and short skip stations start to fade. Finally the band closes.

During the winter months the band does not stay open nearly as long in the summer and to gain the best worldwide conditions the spring and autumn are best. Also at the trough of the sunspot cycle the band may not support long distance contacts for many days. To give an indication of whether the band is likely to be open a quick listen to the 13 metre broadcast band gives a good indication. The presence of a large number of broadcast stations will indicate that there are good prospects of receiving amateur stations. If no broadcast stations are present then there is little likelihood of any amateur stations being heard.

Although beacons do not currently exist for 15 metres, a similar network to that on 14.100 MHz is planned for 21.150 MHz.

8.5.8 12 metres

This is the third of the bands released for amateur operation after the 1979 Conference. Like the 17 metre band it is relatively small and activity is still less than the more established bands either side. Nevertheless it is a good hunting ground and can produce a variety of stations from all over the world.

It is very similar to the bands either side of it, providing a useful half way stage between the two. It remains open for longer in the day than 10 metres, and is slightly less affected by the sunspot cycle. Even so it does not support much activity during the years of the sunspot minimum.

8.5.9 10 metres

This is a band which can bring many surprises. It remains closed during the trough of the sunspot cycle, but at its peak it is possible to hear stations all over the world, even those using very little power. As a result of this and the amount of spectrum bandwidth available, it is possibly one of the most interesting bands to monitor. Even during the years of the sunspot minimum it is possible to hear stations in the summer up to distances of 2000 km as a result of sporadic E.

There is a variety of types of activity. Apart from the usual long distance modes like SSB, and Morse together with the data modes, there is also a growing amount of FM activity located around the top of the band. This has been helped along by the availability of Citizen's Band sets which can often be modified to operate on 10 metres. As most of this operation is of a semi-local nature, radio amateurs in the USA have set up a number of repeaters. When the conditions are good, these can be heard worldwide, and stations from many countries use them.

Beacons are an almost essential feature of this band. As conditions are very variable, it is useful to be able to monitor them to see when the band is open. At the moment there is a large number of single frequency beacons which operate independently of one another on different frequencies. To be able to gain an idea of conditions it is necessary to tune over the beacon sub-band. Although the allocation for beacons is now 28.190 to 28.225 MHz it originally extended up to 28.300 MHz and some beacons can still be heard between 28.225 and 28.300 MHz. For the future there are plans to have a time multiplexed system like that in operation on 20 metres.

These band descriptions can naturally only give a very generalized view of the bands. The only real way to develop a feel for the bands is to spend time listening to them all. Once this has been gained, it is very much easier to judge the best band to use at a particular time of day and to hear the sorts of stations being sought.

8.6 Band plans

To enable radio amateurs to make the most efficient use of the allocations given to them, the bands are split up into different areas where different modes are used. This is done to minimize interference and to enable stations searching for a contact on a particular mode to know where to look.

Usually Morse transmissions are found at the bottom of the band, although it is permissible to use Morse in any part of the band. The reason for this is that Morse is a narrow band mode and will not cause any additional interference over a single sideband transmission. Also if a contact becomes marginal and copy is difficult, it is often necessary to revert to Morse to maintain contact.

The band plan for the UK is given in Table 8.7. This is broadly true for the rest of the world, but in places like the USA where there are larger allocations and the licence conditions are different, this will give rise to some differences. However, for listening it gives the areas where stations are found on a particular mode.

Table 8.7 HF Band Plan

From		To	Use
1.810	–	1.840	CW
1.840	–	2.000	Phone and CW
3.500	–	3.600	CW
3.500	–	3.510	Intercontinental Contacts
3.560			QRP Frequency
3.600	–	3.800	Phone and CW
7.000	–	7.040	CW
7.030			QRP frequency
7.040	–	7.100	Phone and CW
10.100	–	10.140	CW
10.106			QRP Frequency
10.140	–	10.150	CW and RTTY
14.000	–	14.099	CW
14.060			QRP frequency
14.099	–	14.101	Beacons
14.101	–	14.350	Phone and CW
18.068	–	18.100	CW
18.100	–	18.110	CW and RTTY
18.110	–	18.168	Phone and CW
21.000	–	21.149	CW Only
21.060			QRP frequency
21.149	–	21.151	Beacons
21.151	–	21.450	Phone and CW
24.890	–	24.920	CW
24.920	–	24.930	CW and RTTY
24.930	–	24.990	Phone and CW
28.000	–	28.190	CW Only
28.060			QRP Frequency
28.190	–	28.225	Beacons
28.225	–	29.700	Phone and CW

8.7 Language

With stations audible from all over the world on the amateur bands it is possible to hear a wide variety of languages. Fortunately for the English speaking listener, many contacts are in English and this has become the main language on the air. Naturally many contacts are made in other languages. Those from Latin America naturally have a large number of contacts in Spanish and Portuguese. Russian is another widely used language. However for the non-English speaker, the wide use of abbreviations and code helps ensure that it is possible to make contacts with only the minimum knowledge of English. This is particularly true of Morse contacts where abbreviations and codes account for most of what is sent.

8.8 QSL cards

After an interesting contact or one with a station in a country which has not been contacted before it is often nice to have some form of confirmation. To do this radio amateurs often exchange **QSL cards**. The name comes from the Q code meaning 'I confirm receipt' and the cards themselves are usually about postcard sized. Many are very colourful, sometimes with photographs of the station.

The cards contain the basic information about the contact. Time, date, frequency, report, mode of transmission and the equipment are all included as shown in the example in Figure 8.2.

QTH: 5 Fleming Road, Hamstone, England

G3QQQ

Operator: Fred Smith
To Radio ....
This is to confirm our two way AM/SSB/CW
Contact on .... at .... GMT on .... MHz
Your signals were RST ....
Rig .... Antenna ....
73 .... PSE/Tnx QSL Direct/via Bureau

Figure 8.2 *A typical QSL card*

QSL card collecting can be an interesting sideline to the hobby. After a few years on the air hams can easily have collections of several hundred cards. As a good proportion of them are colourful and interesting, they make good decorations. Many hams use them for decorating the radio room or shack, showing the various countries with which they have made contact.

Figure 8.3 *A variety of QSL cards*

Short wave listeners can also collect cards. This can be done by sending reports to stations and asking for a card in return. The reports should be as useful as possible to ensure the best chance of receiving a card in return. It is worth remembering that stations on sparsely populated islands or other countries where the radio amateur population is small can receive large numbers of listener reports on top of the large number of cards confirming contacts. In instances like these it is especially important that the report is made clear, concise and useful.

The report should detail as many details as possible about the signal being received. The date, time, frequency and callsign of the station who was being contacted or whether a CQ call was being made are all essential. A concise summary of the receiving equipment, including the aerial, should be given. A summary of the propagation conditions is also useful, giving examples of other stations or areas which were audible. It also helps to tell the station any relevant information he or she may not already know. For example it may be that the signals are reaching a part of the world he or she was not in contact with. Low power or QRP stations are also more likely to be interested in reports, especially if they have been heard over a considerable distance. If details like these are included or reports of the station being heard over a period of time are included, a QSL card is more likely to be received. If the card is being sent directly through the post, then the return postage must be included. This can be done by using an International Reply Coupon (IRC) which can be bought at a post office.

When receiving QSL cards and checking them off in a logbook, it is worth remembering that in the USA the dates are written slightly differently. In most countries a date is written day/month/year, e.g. 25/12/1995 (25th December 1995). In the USA the date is written month/day/year, e.g. 12/25/1995.

When listening to Morse contacts, sometimes the code QSLL will be heard. This means that the station will return a QSL card to the station with whom he or she is in contact if or when one is received. In other words he or she does not send out cards for all contacts.

8.9 QSL bureau

Sending QSL cards individually through the post can quickly become very expensive. Some stations may want to send out several hundred cards a year for contacts which have been made. For short wave listeners it becomes even more expensive as the return postage has to be included every time if there is to be any chance of a reply.

To reduce costs a QSL bureau system has been devised. It enables stations to send all their cards in bulk, thereby greatly reducing the costs. Each station sends their outgoing cards to their national bureau. In the case of Great Britain the bureau is operated by the Radio Society of Great Britain. Once at the bureau the cards are sorted into countries and, along with cards from other stations, they are again forwarded in bulk to the national bureaux of other countries. At the receiving end the cards are sorted for the individual stations and, when sufficient have been received, they are sent on to individual stations.

Sending a card by this method can take much longer. Typically it may take a year or more to receive a reply from a card which has been sent out. However, the reduced costs more than justify the delay. Obviously if a particular card is needed urgently then it can be sent directly.

Not all stations can make use of a bureau. As there are obviously costs involved for the organization running the service, it is necessary for the station to be a member of the national society organizing the bureau, and in some countries a small charge is levied for each outgoing card.

8.10 Awards

In addition to the fascination of listening to amateur stations from all over the world, many people enjoy the challenge of seeking out stations from countries they have not heard before. To recognize achievements like this as well as many others it is possible to obtain operating awards.

The awards themselves consist of fine looking certificates which are ideal for mounting on the wall and many people frame them.

Most awards are for licensed transmitting amateurs, but there are many awards available solely for listeners and many of the transmitting awards are also available to listeners on a 'heard' basis. Details regularly appear in the amateur radio press. In addition to this, information about them can be obtained from the relevant issuing societies. The Radio Society of Great Britain has a number, including the DX Listeners Century Award which is gained by submitting evidence, i.e. QSL cards, to show that 100 countries in the RSGB's country list have been heard.

Another award called the IARU Region 1 Award is given to licensed amateurs and short wave listeners for making contact or hearing stations in countries which are members of Region 1 of the Region 1 Division of the International Amateur Radio Union (IARU).

Figure 8.4 *An Islands on the Air award (Courtesy of the Radio Society of Great Britain)*

One particularly popular award is the Islands on the Air award. It was started by Geoff Watts, a leading UK short wave listener in the 1960s, and it is now firmly established as a very popular award. The basic award is gained by radio amateurs for submitting proof that they have made contact with 100 different islands or groups of islands.

Endorsements can be gained for additional hundreds of islands or groups. This is an attractive award printed in many colours and makes an attractive addition to any shack wall.

These are only three of the awards which are available. For anyone who enjoys measuring the challenges of short wave listening there is a vast number of awards which can be gained. Details of them appear periodically in the amateur radio magazines.

8.11 DXpeditions

Many radio amateurs as well as listeners enjoy making contacts with or hearing new countries. In view of this, radio expeditions are often mounted so that amateur radio operation can take place from a country where there is either very little amateur radio activity or none at all. These expeditions, often called DXpeditions, attract a lot of publicity in the amateur radio press.

When a DXpedition station appears on the bands, vast numbers of stations call them to get a contact and a pile-up of stations is usually heard around them. This familiar gaggle of noise is often the first tell tale sign that an interesting station is about.

Pile-ups can often be difficult to manage. With so many stations on the same frequency it is not easy to hear exactly who is transmitting at any given time. This can result in even the most polite stations transmitting out of turn. To help alleviate the problem some stations operate on split frequencies, i.e. they transmit on one frequency and listen on another. This makes the pile-up far more manageable because stations wanting to contact the DXpedition should be able to hear them with a minimum of interference. Sometimes the DXpedition will specify a band of frequencies to call on. This spaces out the stations calling, reducing the interference levels and making it much easier to identify callsigns.

8.12 Contests

Another aspect of amateur radio which many people enjoy is participating in contests. Throughout the year a variety of them are organized for the short wave bands, some of which are larger than the others. In general the aim of a contest is to make as many contacts as possible, although the rules vary from one to the next. In certain ones the aim may be to contact as many stations from a particular area or country as possible.

The main contests are summarized in Table 8.8. In the major ones, particularly the ARRL DX, WPX and CQ World Wide contests, the bands erupt with a very large amount of activity. Other smaller contests may make much less impact on the bands but are nevertheless still very popular.

During a contest the length of contacts is kept to a minimum. Usually a report and some form of serial information is all that is sent. In one contest it may be the serial number for the contact whereas others it may be the zone. For the Worked All Europe contest the operator has to send his age (ladies send 00!). Once the information has been sent and confirmed by both stations, the next contact is started as soon as possible to ensure that the maximum number of points is gained.

DXpeditions are often organized by groups of people for some of the larger contests. By operating from a country which is less commonly heard, the station attracts more contacts and gains more points. As a result a wide variety of new and interesting stations can be heard.

Table 8.8 Major Amateur Radio Contests

Contest	*Date*	*Comments*
ARRL DX Contest (CW)	Third full W/E February	Stations contact USA/Canada
ARRL DX Contest (SSB)	First full W/E March	Stations contact USA/Canada
CQ-Worked PrefiXes (WPX)(SSB)	Last full W/E March	Stations contact as many stations as possible. Extra points given for new prefixes contacted
CQ-Worked PrefiXes (WPX) (CW)	Last full W/E May	Stations contact as many other stations as possible Extra points given for new prefixes contacted
CW Field Day (UK) (CW)	Usually first W/E June	British portable stations make as many contacts as possible
All Asia (SSB)	Third full W/E June	Contact stations in Asia
IARU-Radiosport (SSB / CW)	Second full W/E July	Contact as many stations as possible Extra points given for new countries contacted
Worked All Europe-DX (CW)	Second full W/E August	Stations outside Europe contact as many European stations as possible
All Asia (CW)	Last full W/E August	Contact stations in Asia
SSB Field Day (SSB)	First full W/E September	Portable stations make as many contacts as possible
Worked All Europe-DX (SSB)	Second full W/E September	Stations outside Europe to contact as many European stations as possible
CQ-World Wide (SSB)	Last full W/E October	Contact as many stations in as many countries as possible
CQ-World Wide (CW)	Last full W/E November	Contact as many stations in as many countries as possible

For the short wave listener these contests give an ideal opportunity to hear new and interesting countries. Not only are DXpeditions mounted, but stations which are not normally active participate. In addition to this many stations operate around the clock, resulting in countries being heard at different times.

Not everybody enjoys participating in contests. Some avoid them, but judging by the levels of activity many more seem to enjoy at least some involvement.

8.13 Listening skills

After some time listening on the bands it is surprising how much easier it becomes to find the interesting stations. Not only does one develop a knowledge of the bands but also a feel for which stations are worth listening to.

The accent of the operator is naturally a major clue. Sometimes it may be the sound of the signal itself. A weak 'fluttery' signal can mean that the path crosses one of the poles. This can be an indication of a distant station. Distant stations can also be heard just as a band is closing. This is particularly true on the higher frequency bands. Here the local interference is at its least and only stations with low angles of radiation are heard. These are generally from the greater distances.

The tell tale noise of a pile-up is also another indication of a station worth listening for. Likewise any station giving out reports in a contest style fashion may be worth noting. Similarly contests generally bring to life stations who would not normally be on the air. Coupled with this many DXpeditions are usually mounted for the major contests, giving an enormous variety of new and interesting countries to hear.

Whatever type of listening one likes, experience on the amateur bands will improve one's skills and enhance the enjoyment of the hobby.

9 Broadcast listening

Another fascinating area for listeners is on the broadcast bands. Here a great variety of stations can be heard from all over the world. Many people find hours of enjoyment listening to the diversity of programming. There is also the achievement of picking up distant low power stations. Then in times of trouble or war there is the additional interest of receiving stations from the affected areas. This can be particularly absorbing because it relates the hobby to current affairs.

Broadcast stations can be heard on a variety of bands. Most people are familiar with the medium wave stations. Even though these stations are intended for local coverage it is possible to find some long distance stations at times. The majority of long distance broadcast listening takes place on the short wave bands. Here stations from all over the world can be picked up on one band or another at almost any time of day.

9.1 Purpose of broadcast stations

Broadcast stations are set up for a variety of reasons. Stations on the medium wave band are generally used for local entertainment and news. In most cases they are not aimed to be heard outside their own country and do not have an international flavour. The same is true of broadcasts on the VHF FM band. These stations carry a similar type of material to their medium wave counterparts, except that the audio quality is much higher and as a result they generally have a higher music content.

On the short wave bands stations are almost exclusively used for international broadcasting. Because of the nature of propagation at these frequencies, it is unlikely that these stations will achieve much coverage within their own country, unless the skip distance falls within the country. As a result short wave stations are used as flagships for their countries. Often they carry propaganda. This was particularly true in the days of the Cold War. Stations from the West could be heard giving exactly the opposite views to those from the Eastern Bloc.

The strength of propaganda on the air waves is still used to great effect. During times of conflict radio broadcasting is an ideal medium for this. Some of the most recent examples of this occurred in the Gulf War and in the former Yugoslavia. Stations were used to put forwards the views of each side. Iraq employed some very powerful transmitters to reach the rest of the Arab world. Naturally this made the transmitting sites a target for the coalition forces in the bombing campaign. Similarly radio stations were attacked in Croatia and Bosnia to silence them during the times of war there.

Fortunately most programming is far more peaceful in nature. Many stations have sought to give their countries credibility on the international scene by producing good reliable programming and news. The BBC World Service is recognized the world over as possibly the best station. In many countries people listen to the BBC rather than their own local stations, knowing the news will be reliable and unbiased.

Figure 9.1 *A studio at the BBC World Service (courtesy of the BBC)*

There are also many religious stations on the air. One of the most obvious examples is Vatican Radio, broadcasting from the Vatican City itself. However this is not the only station. There are many others. One called HCJB broadcasts from Columbia and regularly puts a good signal into the UK and many other parts of the world. There are many other religious stations from all over the world, including a large number from the USA.

9.2 Medium wave listening

Although not strictly short wave listening, the medium wave band stretching from 0.5265 MHz to 1.6065 MHz can produce some interesting stations. One advantage is that it is possible to listen on this band without any special equipment.

Unlike many of the short wave stations, those on the medium wave band are intended primarily for entertainment. Programmes consist of music, plays, news, chat shows and a variety of other types of entertainment. Generally it is of a local nature, especially for the low powered local stations.

There is a very wide variety of stations which can be heard from low power local stations to the higher powered stations for covering large areas of the country. In the UK for example the BBC operates a number of national networks. There are also independent national networks as well. Transmitters for these run many kilowatts and can be heard over large distances. In addition to these there are many local stations. These are only intended to have a relatively small coverage area. By tuning over the medium wave band it is possible to hear a vast and interesting range of stations.

Figure 9.2 *Aerial system for a local radio station (Courtesy of Premier Radio)*

During the day high powered stations may be heard up to distances of around 200 km. The local low powered stations obviously have a much smaller coverage, although they can often be heard over distances of 100 km. This reduction in coverage is hardly surprising in view of the fact that some only use powers of around 100 watts.

In other countries similar distances may be covered. In the USA there is a large number of stations, and a wide variety of listening.

At dusk the distances over which signals can be heard greatly increase. Interference levels rise as stations from further afield become audible. In view of the very high usage of all the frequencies the secret of listening is to discover the frequencies where local signals are not present. If this can be done then stations up to distances of 1000 km and more can sometimes be heard.

Many countries now use the medium wave band for cross-border broadcasting. The BBC World Service broadcasts to Europe on 648 kHz and Radio Moscow World Service broadcasting with 1.2 megawatts from St Petersburg on 1494 kHz and can easily be heard over much of Northern Europe at distances in excess of 1500 km.

Many other long distance stations can be heard with a little patience. Some North American stations can be heard in Europe after midnight. Sometimes it may also be possible to hear stations from South America, although this is more difficult. Similarly listeners in other parts of the world can hear long distance stations. Those in the North America can often hear stations from the Caribbean and South America as well as those further afield.

Reception of long distance stations is not easy although it is very rewarding. Even when local interference levels allow these stations to be heard, it is often found that two distant stations can be heard on the same channel. Another problem is that splatter from local stations on nearby channels can make reception difficult. When this occurs narrow band filters on the receiver become very useful.

In recent years the broadcast bands have been standardized so that stations transmit on particular channels. The channel spacing on the medium wave band is 9 kHz in Europe and 10 kHz in the USA. This reduces the amount of interference because the number of annoying heterodynes from stations 1 or 2 kilohertz away from one another are eliminated.

One of the main requirements for medium wave band listening is a directional aerial Most portable radios use a ferrite rod aerial, and receivers like these are only really acceptable as an introduction. More sophisticated communications sets are generally required if transatlantic stations are to be sought. These sets need external aerials and a loop aerial like the one described in Chapter 5 is ideal. It is not too large to be unwieldy and gives directivity, allowing unwanted interference to be minimized. In view of the levels of interference the aerial is usually rotated to reduce the unwanted signal rather than to maximize the wanted one, although individual conditions usually govern the best setting.

9.3 Short wave broadcasts

Broadcasting on the short wave bands is very different to that on the medium wave band. Stations from all over the world are heard. Interference levels are also much higher. On the medium wave it is relatively easy to ensure that stations using the same frequency are sufficiently far away from one another not to cause any interference. This is clearly not possible on the short wave bands where signals travel much further and conditions vary between one day and the next. However, international meetings are held between the major broadcasters to try to plan which frequencies they will use in order to keep interference levels to a minimum.

One of the ways of organizing the short wave bands to obtain the maximum usage is to split each band into distinct channels. For short wave broadcasting a spacing of 5 kHz is used. The fact that the spacing is much smaller than that on the medium wave band gives an indication of the pressure on the available band space. In turn this shows the importance which is placed against short wave broadcasting.

As the spacing is much smaller, the transmission quality is not as good as on the medium waves where a wider bandwidth is available. In view of the distortion often caused by propagation of the signal via the ionosphere, quality is normally not a major issue.

Within the short wave spectrum there are a number of bands which are not used for international broadcasting in the normal sense. These bands are called **tropical bands** and are found in the lower portion of the short wave spectrum.

9.4 Tropical bands

Many of the countries in the tropical areas of the world are large and relatively sparsely populated. This means that it is not possible to obtain adequate coverage with either medium wave or VHF FM transmissions. In countries like the UK, 20 or more stations may be required to give reasonable coverage for a national network. In countries which are possibly larger and are less densely populated it is not economically viable to operate as many transmitters. It is, however, just as important to ensure that the population is able to pick up radio broadcasts to keep them aware of the news and national events. Having a radio station also helps to maintain a sense of national identity, a fact which may be a political necessity.

Figure 9.3 *BBC relay station in Hong Kong (Courtesy of the BBC)*

The solution is to use frequencies which enable a greater coverage to be gained. As a result a number of 'tropical bands' are allocated with frequencies up to about 5 MHz. By using the relatively short skip normally present on these bands greater coverage can be achieved. Often transmitters in the medium wave and VHF FM are used for the populated areas whilst the tropical bands are used to give greater coverage for the outlying areas.

Stations using these frequencies are not like the normal short wave broadcasters. Instead they tend to be more like the typical medium wave or VHF FM stations heard in Europe or America. Programmes are national in nature and might include ordinary music programmes, topical discussions and the like. Often the studios are quite rudimentary when compared to some of the medium wave and VHF FM stations found in North America and Europe, but nevertheless they operate an essential service.

Another way in which stations on these bands differ from the large international broadcasters is that the powers are very much lower. Powers of the order of 1 kW are quite common and these are very small when compared to the much larger transmitters of the international broadcast bands. Aerials also tend to be far more modest. Simple dipoles or verticals are often used. These modest powers and aerials are quite adequate as the intention is only to cover the country in question, and not reach vast distances across the world.

The tropical bands are only used by countries between latitudes of 23N and 23S. This area covers part of Africa, Asia, Central and South America.

Listening on these bands can sometimes be difficult. The low transmitter powers mean that signal levels are low. Also the tropical bands suffer from high levels of static. In addition to this the bands are allocated to other services outside the tropical region. This means that interference from utility stations can be high, making listening more difficult.

Language may also be a problem. As the stations broadcast to their own particular country, they will naturally use the language for the indigenous population. This is rarely English, although it can sometimes be heard.

However, when stations are heard they can be very interesting and give a distinct flavour for the country of origin. Coupled to this listening on these bands can stretch the skill of many listeners and this makes stations heard more rewarding and enjoyable.

9.5 Broadcast bands

The broadcast bands are located in all sections of the short wave spectrum as shown in Table 9.1. This means that stations can generally be heard at all times of the day and night. Also by choosing different bands it is possible to change the areas from which stations can be heard. Like the amateur bands, by knowing the characteristics of each band the right band can be chosen for the type of station which is being sought.

Table 9.1 Long, Medium and Short Wave Broadcast Bands

Band	Frequency
Long wave	0.150–0.285
Medium wave	0.5265–1.6065
120 metres	2.300–2.495*
90 metres	3.200–3.400*
75 metres	3.900–4.000†
60 metres	4.750–5.060*
49 metres	5.950–6.200
41 metres	7.100–7.300
31 metres	9.500–9.990
25 metres	11.650–12.050
22 metres	13.600–13.800
19 metres	15.100–15.600
16 metres	17.550–17.900
13 metres	21.450–21.850
11 metres	25.670–26.100

All frequencies are in MHz
* Tropical bands only for use in tropical areas
† Only allocated for broadcasting in Europe and Asia

9.5.1 120 metres

This is the lowest frequency broadcast band in the short wave spectrum. It is one of the tropical bands but one of the least used. Station powers are low and coupled to the difficult propagation for long distance reception, this makes signals very weak when they can be heard.

The band is shared with other services which can be heard in other parts of the world. This too makes listening more difficult.

To receive any stations, a good receiver and aerial are essential and propagation dictates that broadcast stations are only audible after dark in non-tropical areas of the world.

9.5.2 90 metres

This is another tropical band, but far better for listeners than 120 metres. Transmitting stations use much higher powers, some as high as 100 kW, together with better aerials. In addition to this, propagation on these frequencies enables signals to be heard better over longer distances.

Stations are generally only audible after dark and the best season for listening is in the winter months. Conditions are also better during the years of the sunspot minimum. When conditions are good, a variety of stations may be audible from many parts of the tropical region, although again a good aerial and receiver are advised. Again interference from utility stations may be a problem.

9.5.3 75 metres

This band is used mainly by stations in Asia and Africa. However some European stations can be found in the top section of the band, notably the BBC World Service and a number of others including the Voice of America.

One of the main problems with the band is that interference levels are very high. Not only is it shared by other utility stations, but it is used by radio amateurs in some areas of the world. Even though their powers are low when compared to broadcast stations, they can cause interference at times.

Like other bands in this portion of the spectrum, it performs best after dark and in periods of the sunspot minimum.

9.5.4 60 metres

This is the highest in frequency of the tropical bands and it is the best for long distance listening. It is well populated with stations, many of which are very high powered. However there are many low powered ones which make interesting listening.

Propagation on the band is also better for long distance signals. Being higher in frequency than the other tropical bands, it is open for long distance signals for longer. Often it is open in the day, especially towards the end of the afternoon when African stations can be heard in Europe. After dark stations further to the west can be heard. In Europe, South and Central American stations can often be heard with relative ease.

9.5.5 49 metres

This is one of the busiest of the broadcast bands and stations can be heard day and night. Most of the large high powered international stations can be heard on this band at one time or another. This means that the band is very crowded and it is often difficult to hear the low powered stations or those further afield. Often the crowding is made worse during the periods of the sunspot minimum when it becomes very popular with the broadcasters. This is because many of the higher frequency bands are closed at night whilst this one enables much of the world to be covered.

During the day stations within 2000 or 3000 km are generally heard. At night it is usually possible to hear more distant stations because the propagation changes so that the more local stations are not heard.

9.5.6 41 metres

The 41 metre band is only allocated for broadcasting in Europe and Asia. In other parts of the world it is allocated to radio amateurs. Despite the very high powers that are used by the broadcast stations it is often possible to hear radio amateurs from the USA in Europe. Similarly outside Europe broadcast stations can often be heard amidst the amateur radio activity.

In terms of broadcasting the band is widely used, carrying stations within 2000 or 3000 km by day and opening up for longer distance communication at night. However, like the 49 metre band, many of the more distant stations are masked by the high power stations much nearer, particularly during the day. For this reason the band is not favoured by many listeners who prefer bands which have the prospect of hearing lower powered stations from further afield.

9.5.7 31 metres

Being higher in frequency, this band is more affected by the sunspot cycle than those lower in frequency. It is usually open around the clock, particularly in the periods of the sunspot maximum, and can produce some good long haul stations at night, although dawn and dusk can produce the best results. During the day the shorter range signals are heard.

9.5.8 25 metres

The character of this band is very much like 31 metres, although being higher in frequency it is more affected by the sunspot cycle. Here again the band is usually open for most of the day, closing at night in the winter and during the period of the sunspot minimum.

The band is popular with broadcasters and this means that interference levels are usually high. However it is still a good hunting ground for long haul stations. The best times are usually around dawn and dusk, and during spring and autumn. At these times it is often possible to hear station over distances of many thousands of kilometres.

9.5.9 22 metres

The 22 metre band was allocated to broadcasters after the 1979 World Administrative Conference. Accordingly it may not be marked on all short wave receivers. Although a relatively recent allocation, this band is very popular with broadcasters and listeners as well. In terms of propagation it bears many similarities to the bands either side of it. Stations are usually audible throughout most of the day and night, although in times of the sunspot minimum and during the winter it will close in the evening.

During the day stations between about 1000 and 3000 km are generally audible. After dark this range extends considerably and stations over 1500 km to anywhere around the world should be audible.

9.5.10 19 metres

The character of this band is very dependent upon the state of the ionosphere. In the trough of the sunspot cycle, conditions on this band are not nearly as good as at the peak. However, it is normally possible to hear a wide variety of stations during the day. Possibilities for distant stations are generally best around dawn and dusk, although there are possibilities at any time. Again spring and autumn usually produce the best conditions.

In view of the nature of the band a wide variety of broadcasters are present. It is possible to hear distant stations using relatively low powers, although there is always strong competition from the larger stations.

9.5.11 16 metres

With the frequency rising, the short range stations are less prominent. During the hours of daylight it is often possible to hear some distant stations, especially at either end of the day. After dark the band often stays open for several hours, although it is likely to close early in winter and around the sunspot minimum. In view of the greater possibility of long distance stations this band is quite popular with listeners and broadcasters alike.

9.5.12 13 metres

This band is generally considered to be a daytime band, especially in the times of the sunspot minimum. Even then it is less reliable than bands which are lower in frequency because it is very dependent upon the changes in the ionosphere. In periods of sunspot maxima it is open for much longer; it can then remain open well into the night.

When the band is open, it is capable of producing stations from considerable distances. Often low powered stations can be heard from all over the world. However congestion can be high at times due to the long ranges which can be experienced.

9.5.13 11 metres

This is the highest in frequency of the short wave broadcast bands. As such it the least reliable, dependent very much upon the state of the ionosphere. Even at the peak of the sunspot cycle it is considered to be a daytime band. However, it is capable of producing some very good results.

During periods of the sunspot minimum many broadcast stations do not use it at all. The reason is that it is often completely closed, although in the summer there may be a change of propagation via sporadic E. Even the chance of this is not sufficient for broadcasters to use it for their regular transmissions. This means that the band is often devoid of signals for many days or weeks during the years of the sunspot minimum. In view of the relatively low usage of this band its size was reduced at a recent World Administrative Conference. The original allocation extended down to 25.600 MHz.

Broadcasters are allocated specific bands and should keep within them to ensure they do not cause interference to other services. However, when bands become congested they often move to frequencies either side of the agreed allocations. This means that it is worth listening to frequencies either side of the main bands. Here the congestion is much less, and it is possible to pick up a weak long distance station when the band is particularly crowded.

9.6 Jammers

Certain countries do not want listeners in their country to hear certain outside broadcasts. When this happens, broadcasts are sometimes jammed. This was particularly prevalent during the Cold War. To jam a signal, another broadcast transmitter may be tuned to the same frequency. Another more common way is to transmit a specific jamming signal. This consists of a distinctive rasping sound, through which very little can be heard.

To effectively jam a signal over a wide area is very difficult on the short wave bands. Propagation means that the jammer will be heard in places where the signal to be jammed is not audible and conversely the jamming signal may not be audible everywhere the signal to be jammed is heard. Nevertheless many very high power jammers are in use even today on the short wave broadcast bands.

9.7 Relay stations

It is not possible to broadcast effectively to all parts of the world from one location even with high power transmitters and effective aerial systems. Competition from other broadcasters is very high and propagation conditions mean that the long distance paths are not totally reliable and are confined to specific periods of the day.

In order to ensure the required coverage, additional transmitters may be located in other parts of the world. This means that broadcasts from the BBC or Voice of America may not emanate from the country in question. Both of the stations have a large number of relays around the world. Other stations too use a number of relays. Deutsche Welle from Germany has relays in Rwanda, Portugal, Canada, Malta and Sri Lanka, and Radio Nederland uses stations in the Netherlands Antilles and Madagascar.

It is also not uncommon to find stations sharing facilities. The Voice of America is relayed from Wooferton in England, the site of a BBC World Service station.

Originally relay stations operated by receiving short wave broadcasts on one frequency and then transmitting them on another. This obviously meant that the quality was not particularly good. However, when satellites were introduced, broadcasters soon switched over to this method as it provided a far more reliable and satisfactory method of producing a programme feed.

In today's political climate a number of stations in Russia and other areas of the former USSR are available for use. Several western broadcasters have taken the opportunity up as it enables them to broadcast more effectively into a number of new areas of the world.

9.8 Languages

A great variety of languages are used on the airwaves. As broadcasts are beamed to particular areas of the world and target particular audiences, it is natural to use their native language. Even though English is the most widely used, very few people in remote corners of the world will understand it. Accordingly almost every language under the sun can be heard. However, for the English speaker there are still plenty of broadcasts which can be heard and understood. Even if another language is used, it is still possible to decipher the station name on a number of occasions.

9.9 Station identification

There are a number of ways in which stations identify themselves. One of the most common is to use the station name. In the UK the BBC domestic services use the names Radio 1, Radio 2 through to Radio 5. Other local and national services also use this method in the UK. Many short wave services also have a station name. Again the name of the

BBC World Service is known all over the world. Other famous stations use names including the Voice of America, Radio Canada International and so forth.

Some stations, however, use a callsign. One famous short wave station is HCJB. In the USA and a number of other countries even the local stations on the long, medium and VHF FM bands use callsigns. Where callsigns are used, they fall into line with the internationally agreed callsign prefix allocations given in the appendix. From this it can be seen that HCJB is based in Columbia and WNYW is in the USA.

9.10 Interval signals

Many stations use what are called **interval signals**. These are short tunes which can be played during short intervals, normally just before the hour when a news or other programme will start. The tunes are usually quite short and very distinctive. The BBC World Service uses the familiar *Lillibulero* which is longer than many. Other shorter tunes may be repeated several times. By using these tunes listeners are quickly able to identify a station and tune into the required one.

9.11 Programme material

The types of programme which are broadcast depend very much upon the station and the intended audience. Typically they are heavily news orientated and most programmes are relatively short. Often they will be comment or discussion programmes, especially following a news bulletin. Factual or documentary style programmes about the country where the station is located are popular. Often they give details about the way of life, a particular area of the country or tell of its history. These programmes can be very interesting as they give a good insight into life in other areas of the world.

Music does not usually feature as highly as it does on domestic services. This is because the quality of transmission over the short waves is not very high, although pop type music is more common. Classical music also suffers the disadvantage that it can be relatively long, preventing it from fitting within the short programme framework of many stations. Even so it may be heard on occasions.

Most of the large international broadcasters are supported by their governments. However, many stations receive little funding from outside. Accordingly some stations include advertisements in their programming. Often these are totally different to the usual style encountered in Europe. Those from Latin America may appear suddenly in the middle of a programme, with no specific break for them. Even some large broadcasters carry commercials. Broadcasts from Russia have been known to have some in order to raise extra finance.

9.12 Schedules

Stations change their frequencies and programme schedules. They do this to make the best use of the propagation. As conditions vary according to the different points in the sunspot cycle, time of year and so forth, the optimum frequencies change. As a result most stations change their schedules, often in the autumn and spring.

Many stations publish listening schedules giving details of frequencies and programmes. There are many books published which give details of the frequencies used by international broadcasters. These are an indispensable aid for the listener and most are updated each year. When buying one check that it is regularly updated and it is current. However without any of these it is still possible to keep track of many stations. They often tend to keep the same frequencies although times and transmitters are more likely to change.

9.13 Short wave listeners' programmes

A number of stations produce programmes which are designed specifically for short wave listeners. They cover a wide variety of topics, including items like receiver reviews, listeners letters, new developments in the field of radio and a host of topics of general interest to those interested in listening as a hobby. As the style and content of these broadcasts vary quite considerably, it is worth listening to a few to decide which is the most interesting.

9.14 QSLing

Collecting QSL cards from broadcast stations can be a fascinating sideline to the hobby in the same way that it is for amateur radio. Most of the large international broadcasters will reply to listener reports and many of the smaller ones are glad to receive informative reports about the reception of their signals.

Some of the very large broadcasters already know how well their signals are received around the world and listener reports are of little use. In view of this and the large cost involved in replying to all the reports which would come in, the BBC no longer gives QSL cards out. It is likely that many more of the large stations will follow this in the years to come as technology improves and the need to rely on listener reports diminishes.

Many stations still welcome listener reports. Not only does it give an indication of how the signal is being received, it also provides useful market research, indicating levels of listeners in different areas of the world. As a result many stations regularly give out their addresses and the requirements for reports.

Reports should be carefully compiled over a period of time. Date, time and frequency are of paramount importance and also include the signal strength, propagation, interference levels and strength of other stations from the area. It can also help to include whether the broadcasts can be heard on any other frequencies. Also do not forget to include several points about the programmes being transmitted as stations usually require this to verify that their signals have actually been heard.

When sending off the report, it helps to include the return postage in the form of international reply coupons (available from the post office) as the cost of QSLs and postage can become very high and a drain on the resources of many stations.

9.15 Major stations

Most countries have their own short wave broadcast stations. Even some of the smaller countries which might appear to have no need of an international station will have some form of short wave station. Many use powers of only 50 kW or less, and may only have a small studio, but they are still capable of putting a presence onto the short wave bands.

In comparison a number of the major stations have sophisticated studios, control and distribution networks and have some very powerful transmitters coupled to efficient aerial systems. Many are very famous, like the BBC World Service.

Based at Bush House in London, the BBC World Service has been in existence since 1932. Its initial purpose was to ensure that every part of the British Empire was able to hear a broadcast from London at some time each day. Since then its main aim has changed. It is now an ambassador for the UK, and it has gained an international reputation for reliable reporting and a high standard of programming.

Programmes are broadcast in over 35 languages to all corners of the earth. Listening figures for the BBC show that 120 million people around the world listen. This does not include countries like China where estimates cannot be made and so the real listening figures are much larger.

To reach these audiences the BBC has a large network of transmitters and relays. In the UK there are stations at Rampisham in Dorset, Skelton in Cumbria, Wooferton in Shropshire and Ordfordness (648 kHz medium wave only). Daventry, the birthplace of the World Service, was closed at the end of 1992 after a long and distinguished service.

There are many relay stations around the world: Ascension Island in the Atlantic, Singapore, Cyprus, Masirah in the Gulf, Hong Kong, Antigua and the Seychelles. In addition to this there are other transmitters in Lesotho and Berlin and exchange facilities in Northern America at Sackville (New Brunswick), Bethany (Ohio), Delano (California) and Greenville (North Carolina).

At these sites a variety of transmitters or senders are used. Most are either 100 kW or 250 kW, although some are 500 kW and some at Masirah are 750 kW. The aerial arrays are equally important. Many of them are directive with control of azimuth and elevation to ensure the maximum signal reaches its destination.

The former Soviet Union is finding it does not need all the transmitters it used to use. Some of these are being used by international broadcasters to reach new areas of the globe more easily. The BBC is no exception and is developing these possibilities.

In addition to all of these the World Service appears on a number of satellite channels and a number of local radio stations, especially those in the Caribbean and the USA take feeds from them to use in their own broadcasts. This is a new and developing area of their market.

Figure 9.4 *The control room for the BBC World Service in Bush House, London (Courtesy of the BBC)*

Bush House is the nerve centre of the whole organization. Programmes are generated here. The news room uses the latest in technology to ensure that the bulletins can be put together efficiently, taking feeds from news agencies and there are also direct links into the other BBC domestic services. For the programmes themselves there is a large number of studios. These are used for everything from preparing recorded programmes to live news broadcasts.

To control the whole operation there is a comprehensive control room. This ensures that the correct programmes are put out over the correct routes to the required sites and transmitters. At any given time there will be programmes leaving Bush House in several different languages destined for transmitter sites all over the world. This is a complicated operation and as a result any schedule changes require to be well planned to ensure that the correct transmitters receive the correct programme feed at the correct time.

9.16 Developments in broadcasting

The field of international broadcasting is very competitive. All broadcasters are seeking to increase the numbers of listeners. Often this is achieved by ensuring that they have satisfactory coverage of all the areas in question. They may increase the number of languages they use or improve the programming standards.

However, many broadcasters are looking to see how the latest technology can help. For example, many international stations can be received via satellite like the BBC World Service.

Other methods are being sought to improve reception on the short waves. Currently broadcasters use ordinary amplitude modulation. This is not very efficient in its use of the spectrum which is very crowded. To help alleviate this it is planned that broadcasts will start to switch to use single sideband. Whilst it is realized that this cannot occur overnight, it is planned that a switch to this mode will be made a number of years into the twenty-first century. The exact date of the changeover has recently been deferred. When broadcast stations use SSB, they will only transmit one sideband. Also the level of the carrier will be reduced by 6 dB which will enable major savings to be made in the power required for the transmitter.

One of the problems with short wave reception is that it is difficult to know what other frequencies are available to receive the station. This type of facility is available on the VHF FM band with the radio data system (RDS). Experiments are now being undertaken to investigate a similar system for AM. Called the AM data system (AMDS) a few stations including the BBC World Service and Deutsche Welle have been evaluating it for use on their short wave transmissions.

10 Setting up a station

Everyone has a different idea of where to place the receiver. For some the receiver may be a portable set and brought out only when it is required for use. Other sets may be kept in the corner of a room and used only occasionally. Many people will want to have a specific area for a radio room or 'shack'. This may house a large array of radio equipment and possibly even a computer as well.

10.1 Location of the shack

Wherever the equipment is placed, there are a number of basic requirements. The area should be comfortable, so that the maximum enjoyment can be gained from the hobby. The convenience, temperature and conditions such as dampness should all be considered. They all play a major part in determining how easy and pleasant it is to operate the equipment and listen on the bands.

Another point to note is that the equipment should not be in a place where it will cause annoyance to others when it is being used. Conversely the set should be in a place where the noise levels do not make it difficult to hear what is coming from the set.

Access for aerial cables is obviously a necessity if serious listening is envisaged and external aerials are required. Some coaxial cables may be quite thick. Often low loss cables may have an outside diameter of 10 or 12 mm. This means that any area for the equipment must have reasonable access for cables of this nature. Similarly there must be an adequate mains supply.

There is a very wide variety of places where the equipment can be placed. The ideal solution for many is to be able to devote a complete room to the hobby. Unfortunately this is not always possible and other options have to be found. However, with a little ingenuity and some thought, quite a number of areas around the home can be pressed into service.

Figure 10.1 *A radio shack using a spare room in a house*

A careful look around the house may reveal a number of suitable places. Many people have successfully installed their equipment in the loft or attic. This has the advantage that this area is rarely used for anything other than storage. A little tidying up can reveal a useful amount of space. However, care must be taken to ensure that the joists are sufficiently strong to take any additional load. It also suffers from the disadvantage that the temperature varies quite considerably, becoming cold in winter and hot in the summer.

Many people have pressed an old garden shed into use as a shack. This solution has many advantages. With a little work a shed can be made into a very comfortable room for listening. The inside walls can be lined with hardboard or some other suitable material to keep it warm, and the inside could even be painted. It also has ideal access for aerial feeders, and it does not take too much effort to run a mains cable in. One of the major disadvantages of this type of shack is the lack of security. A shed with expensive equipment could make an easy target for a thief, although there are many good burglar alarms available today and one of these could be easily installed.

Often a space in a garage can be converted into a suitable radio shack. This may take a reasonable amount of work if the shack is to be made comfortable. Even if a limited amount of work is to be undertaken, steps will have to be taken to eliminate draughts as these may be particularly noticeable in winter.

With a little ingenuity some people have used cupboards very successfully. Particularly the walk-in variety can be ideal, even if they are rather compact. Access for mains and aerial feeders may prove to be difficult in some cases. The advantage of a large cupboard is that it is within the main house and the problems of temperature and security are overcome.

The ideal solution for many people is to set aside a room within the house for the hobby. Often this will be a spare bedroom. This gives easy access and it is possible to drop into the shack to monitor the state of the bands periodically. A room of this size is more likely to give sufficient space for all the equipment to be housed. Mains power will also be available and access for feeders is normally not a great problem, especially if feeders can be routed through the roof space into the room. Some aerials may be housed in the roof space, whereas any outside aerials can have the feeders come directly in through the wall or taken through the roof as required.

10.2 Table

The installation of the correct type of table can make a great improvement to the room. It should be sufficiently large as it should be capable of holding all the equipment. Many stations consist of more than the basic receiver. Other items present may include aerial tuning units, additional receivers or scanners, and, for many people today, a computer. All these items take a significant amount of desk or table space, and the weight soon mounts up. If the surface area is insufficient it can lead to conditions becoming cramped.

The width of the work surface should be sufficient to accommodate all the equipment which is envisaged to be in the shack, although dimensions in the room may dictate how long it can be. Depth is also important. There should be room behind the set for cables, remembering that some coaxial cables do not bend easily and will require more space behind the set. There should also be sufficient in front to the set to enable easy operation. Books such as frequency listings, and a logbook if one is used all need to be spaced out. Ideally about 12 to 18 inches should be available in front of the set. This will be very useful if any constructional projects are to be undertaken. It will give sufficient room for working, allowing the set to be left on whilst the project is under way.

Often a table surface will be fitted to the walls. When doing this it is necessary to leave sufficient room for cables. It is best to leave enough room to pass cables through with their connectors fitted. Normally the largest connectors are mains plugs. These are likely to be moved fairly frequently as new equipment is installed and old equipment is removed. It can be annoying to have to remove the mains plugs each time.

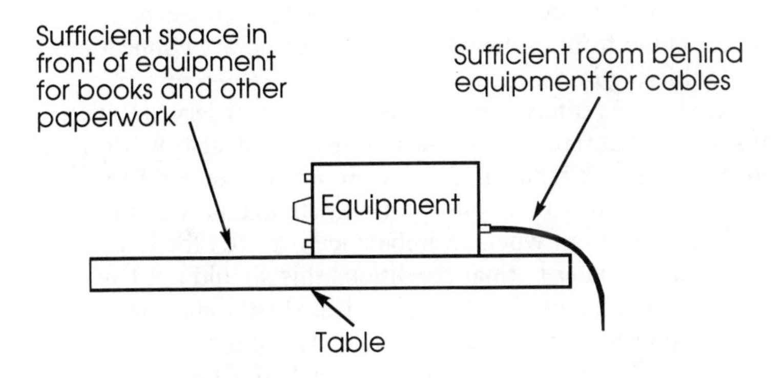

Figure 10.2 *Table top spacing*

10.3 Lighting

It is very important to ensure that the lighting is adequate, particularly if any construction work is envisaged. A dimly light room will quickly prove to be a limitation. Ideally the room or area should have a good overhead light. This should be carefully positioned so that it does not give a large area of shadow over the work space.

An anglepoise lamp or other form of localized light is a great help. A light of this nature is ideal because it can be brought on to the required area.

Incandescent lamps are best for the shack. Fluorescent varieties may appear to be attractive, but they can create high levels of wideband noise which extends up into the microwave regions of the radio spectrum. Even though the receiving aerials may be outside the shack, some interference may still be picked up.

10.4 Electrics

The mains wiring should be considered when all the other parts of the set-up are being put together. It is very easy to forget about any wiring and have to put it in at a later date. As many shacks will have several pieces of mains-powered equipment, the ordinary mains supply would need some multiway adaptors to be used. These can be very inconvenient.

One way of overcoming the problem of the number of sockets required is to fix one or more of the multiway socket strips underneath the back of the table. In this way the mains cables can be neatly run from the back of the equipment and down behind the table, keeping them out of the way. If necessary some provision can be added for providing power at the front of the table for any temporary connections or for items like soldering irons.

At all times safety precautions should be observed with the wiring and fuses of the correct rating must be used. It is also worth considering the installation of a residual current circuit breaker or RCCB, especially if any construction or experimentation with mains equipment is considered.

RCCBs trip out when an imbalance between the live and neutral lines is detected. Under normal conditions this should not happen. Only when a fault occurs should the breaker trip. Installing one of these breakers will help protect the installation and ensure that it is as safe as possible. However, it should be remembered that RCCBs do not prevent electric shocks; they only help reduce the severity of them by tripping out as soon as any malfunction is detected. A variety of RCCBs are available these days and they can easily be installed in the shack wiring.

10.5 Buying a receiver

As the receiver is the main piece of equipment in any short wave listening station, it is worth spending some time exploring exactly what is required. There is a very wide variety of equipment available and sometimes it can be quite bewildering trying to make a choice.

Before going out and parting with any hard earned money it is worth spending some time to decide exactly what is wanted. Many of the short wave listener and amateur radio magazines carry reviews of different sets which can give a good view of what they are like. The advertisements in these magazines also show what is available.

When considering the type of receiver required, look at the facilities which are likely to be used. Frequency range is obviously important. Most short wave receivers cover down to at least 2 MHz, but if medium and long wave broadcast reception is required then lower frequency coverage is required. Many sets will actually tune down to a few hundred kilohertz, but it is worth looking at the sensitivity specifications for these frequencies. Usually they are much reduced when compared to their performance at higher frequencies.

The top frequency is also important. Most short wave sets cover up to 29.999 MHz and this is quite adequate for short wave applications. Higher frequencies are normally only covered by scanners which have coverages which extend up to several hundred megahertz.

The main filters need to be considered. Different bandwidths are needed for different modes. For short wave broadcast AM reception 6 kHz is normally used, 2.7 kHz or thereabouts for SSB, and narrower filters, often 500 Hz or even 250 Hz for Morse, although the normal SSB filter can suffice in many instances.

Another feature which is increasingly important on sets like scanners or the 'World Band' type portable sets is the tuning step size. Many of these sets will only tune in increments of 1 kHz and in the cases of some scanners it may be 5 kHz. This may be satisfactory for broadcast listening where stations on the short wavebands are spaced at 5 kHz intervals, but for many other types of transmission it is unlikely to be suitable.

These are some of the more fundamental points to be investigated. Obviously the general specification for the receiver should be examined, and compared with other sets in the allocated price band.

Apart from the performance of the set itself, it is necessary to consider whether to buy a new set or a secondhand one. It is obviously nice to buy a new set. If this approach is adopted, the latest technology should be assured. There are also unlikely to be any faults with the set and if it is well treated it should give many years of trouble-free performance. However, it is possible to pick up some very good bargains on the secondhand market if a little care is taken.

Obviously there are more risks involved in buying a secondhand set. Those with some experience can usually sort a good set from a bad one. Try it out and see how it performs. It is not always easy to give a set a complete performance assessment, but often it is possible to pick up whether there are any major problems. Also look to see if there are signs of excessive wear and whether the set has been well looked after. These should all be taken into account when making a decision.

Secondhand receivers do not have to be bought privately. Many dealers have a good selection and a reputable dealer is unlikely to sell a set which is not up to standard. Most dealers offer a guarantee period so that if any problems do occur then they will be fixed, but check the terms of the sale before buying.

10.6 Receiver ancillaries

There are many pieces of equipment which can be added on to the receiver as described in Chapter 6. Like the receiver, consideration can be given to buying second hand or new. Again bargains can be obtained quite easily.

One point to note when buying aerial tuning units is that many are designed and manufactured for amateur radio transmitting stations. As the transmitting units need components which can withstand much higher powers they are considerably more expensive. If the tuning unit is to be used only for receiving, a much cheaper receive-only unit can be bought. Often the cost of a receiving tuning unit is much less than half the cost of a transmitting one. They are also quite easy to build and make an ideal introduction to constructing one's own equipment.

10.7 Additional equipment

Apart from the ancillary equipment to add on to the receiver, there is a wide variety of additional equipment which can be bought for a short wave listening station. Whilst many do not directly help with the receiving of stations, they enable the station to run in a more effective manner.

Computers are becoming increasingly common in short wave listening, finding an increasing number of uses. Obviously they can be used for a wide variety of tasks from data communications to logging the stations received. Software is widely available to perform a number of tasks and much is good value.

However it is worth remembering that some computers will create interference and this can be picked up even if good coaxial cable is used in the shack near the computer.

An invaluable piece of equipment for any short wave listener is a test meter or multimeter. There are bound to be times when even a limited amount of testing may be required. As meters can be bought very cheaply, one can be considered for almost any short wave station.

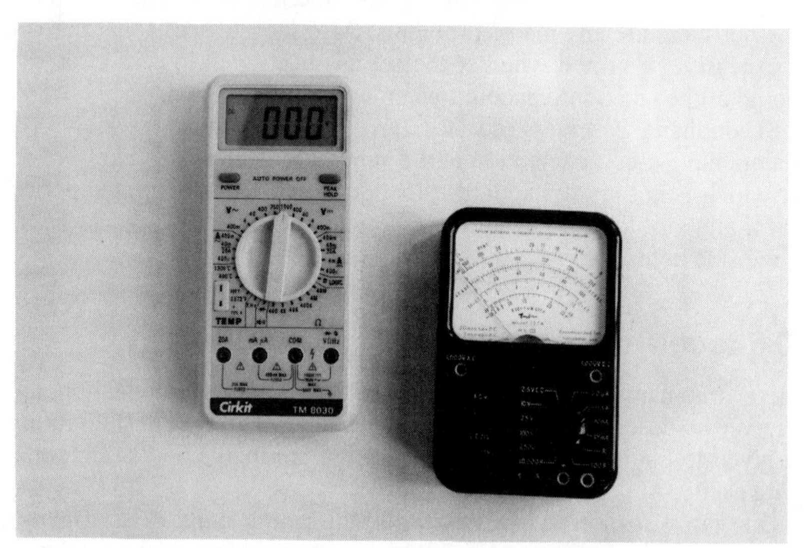

Figure 10.3 *Analogue and digital meters*

There are two basic types of meter: analogue and digital. Analogue ones use meters where a needle and the reading is given by the deflection of the pointer. Digital ones give an numerical readout of the value being monitored. The choice of which type of meter is a matter of personal preference. Analogue ones are ideal for obtaining the trend or approximate value at a glance. The meter itself generally settles quite quickly and the reading is accurate enough for most uses. Normally a view of whether a voltage is present or whether it is in the correct region is all that is needed, and most analogue meters are more than accurate enough.

Digital meters are now more widely available. They offer much greater accuracy and they often have many more functions. It is not uncommon to find frequency and capacitance ranges. Some even measure the temperature or have transistor testing functions. Another advantage is that the impedance of the meter is much higher and as a result they affect the circuit under test far less. Now that their cost is comparable with most analogue meters, digital ones are often the ideal choice.

Another item of equipment which is often useful for a short wave listener is a dip meter. These meters are used for setting up and adjusting aerials for optimum performance amongst other uses. If a large amount of aerial experimentation is envisaged then they can be a worthwhile purchase. However they are relatively specialized and normally only available from amateur radio stockists.

Figure 10.4 *A dip meter*

10.8 Paperwork

Many listeners will want to keep a record of the stations they have heard. This can be useful if it is necessary to see details of a station like the frequency on which it was heard. It is also helpful in recording the reception details of a broadcast station because these stations often prefer reports of reception over a period of time.

Often a logbook may be used to record these details. They can be bought, or alternatively made very easily, particularly if a computer and printer are available. A typical page for a logbook is shown in Figure 10.5, although this can be altered to suit.

Date	Time	Station	In contact with	Freq (MHz)	Report RST	QSL sent	QSL received	Comments

Figure 10.5 *A page from a short wave listener logbook*

Appendix A
ITU callsign prefix allocations

Prefix allocation From	*to*	*Country*
A2A	A2Z	Botswana
A3A	A3Z	Tonga
A4A	A4Z	Oman
A5A	A5Z	Bhutan
A6A	A6Z	United Arab Emirates
A7A	A7Z	Qatar
A8A	A8Z	Liberia
A9A	A9Z	Bahrain
AAA	ALZ	USA
AMA	AOZ	Spain
APA	ASZ	Pakistan
ATA	AWZ	India
AXA	AXZ	Australia
AYA	AZZ	Argentina
BAA	BZZ	China
C2A	C2Z	Nauru
C3A	C3Z	Andorra
C4A	C4Z	Cyprus
C5A	C5Z	The Gambia
C6A	C6Z	Bahamas
C7A	C7Z	World Meteorological Organisation
C8A	C9Z	Mozambique
CAA	CEZ	Chile
CFA	CKZ	Canada
CLA	CMZ	Cuba
CNA	CNZ	Morocco
COA	COZ	Cuba
CPA	CPZ	Bolivia
CQA	CUZ	Portugal
CVA	CXZ	Uraguay
CYA	CZZ	Canada
D2A	D3Z	Angola
D4A	D4Z	Cape Verde Is
D5A	D5Z	Liberia
D6A	D6Z	Comoros Is
D7A	D9Z	Republic of Korea (South Korea)
DAA	DRZ	Germany
DSA	DTZ	Republic of Korea (South Korea)
DUA	DZZ	Philippines
E2A	E2Z	Thailand
E3A	E3Z	Eritrea
EAA	EHZ	Spain
EIA	EJZ	Eire
EKA	EKZ	Armenia
ELA	ELZ	Liberia
EMA	EOZ	Ukraine
EPA	EQZ	Iran
ERA	ERZ	Moldova
ESA	ESZ	Estonia

continued overleaf

continued from previous page

Prefix allocation From	*to*	*Country*
ETA	ETZ	Ethiopia
EUA	EWZ	Belarus
EXA	EXZ	Kyrghyzstan
EYA	EYZ	Tadjikistan
EZA	EZZ	Turkmenistan
FAA	FZZ	France
GAA	GZZ	United Kingdom
H2A	H2Z	Cyprus
H3A	H3Z	Panama
H4A	H4Z	Solomon Is
H5A	H5Z	Bophuthatswana
H6A	H7Z	Nicaragua
H8A	H9Z	Panama
HAA	HAZ	Hungary
HBA	HBZ	Switzerland
HCA	HDZ	Ecuador
HEA	HEZ	Switzerland
HFA	HFZ	Poland
HGA	HGZ	Hungary
HHA	HHZ	Haiti
HIA	HIZ	Dominican Republic
HJA	HKZ	Colombia
HLA	HLZ	Republic of Korea (South Korea)
HMA	HMZ	Democratic People's Republic of Korea (North Korea)
HNA	HNZ	Iraq
HOA	HPZ	Panama
HQA	HRZ	Honduras
HSA	HSZ	Thailand
HTA	HTZ	Nicaragua
HUA	HUZ	El Salvador
HVA	HVZ	Vatican
HWA	HYZ	France
HZA	HZZ	Saudi Arabia
IAA	IZZ	Italy
J2A	J2Z	Djibouti Republic
J3A	J3Z	Grenada
J4A	J4Z	Greece
J5A	J5Z	Guinea Bissau
J6A	J6Z	St Lucia
J7A	J7Z	Dominica
J8A	J8Z	St Vincent & Grenadines
JAA	JSZ	Japan
JTA	JVZ	Mongolia
JWA	JXZ	Norway
JYA	JYZ	Jordan
JZA	JZZ	Indonesia
KAA	KZZ	USA
L2A	L9Z	Argentina
LAA	LNZ	Norway
LOA	LWZ	Argentina
LXA	LXZ	Luxembourg
LYA	LYZ	Lithuania
LZA	LZZ	Bulgaria
MAA	MZZ	United Kingdom
NAA	NZZ	USA
OAA	OCZ	Peru
ODA	ODZ	Lebanon
OEA	OEZ	Austria
OFA	OJZ	Finland
OKA	OLZ	Czech Republic
OMA	OMZ	Slovak Republic
ONA	OTZ	Belgium
OUA	OZZ	Denmark
P2A	P2Z	Papua New Guinea
P3A	P3Z	Cyprus
P4A	P4Z	Aruba
P5A	P9Z	Democratic People's Republic of Korea (North Korea)
PAA	PIZ	Netherlands
PJA	PJZ	Netherlands Antilles
PKA	POZ	Indonesia
PPA	PYZ	Brazil
PZA	PZZ	Suriname
RAA	RZZ	Russian Federation
S2A	S3Z	Bangladesh
S5A	S5Z	Slovenia
S6A	S6Z	Singapore
S7A	S7Z	Seychelles
S8A	S8Z	Transkei
S9A	S9Z	Sao Tome and Principe
SAA	SMZ	Sweden
SNA	SRZ	Poland
SSA	SSM	Egypt
SSN	STZ	Sudan
SUA	SUZ	Egypt
SVA	SZZ	Greece
T2A	T2Z	Tuvalu
T3A	T3Z	Kiribati
T4A	T4Z	Cuba
T5A	T5Z	Somali Republic
T6A	T6Z	Afghanistan

continued overleaf

continued from previous page

Prefix allocation From	*to*	*Country*
T7A	T7Z	San Marino
T9A	T9Z	Bosnia Herzegovina
TAA	TCZ	Turkey
TDA	TDZ	Guatemala
TEA	TEZ	Costa Rica
TFA	TFZ	Iceland
TGA	TGZ	Guatemala
THA	THZ	France
TIA	TIZ	Costa Rica
TJA	TJZ	Cameroon
TKA	TKZ	France
TLA	TLZ	Central African Republic
TMA	TMZ	France
TNA	TNZ	Congo
TOA	TQZ	France
TRA	TRZ	Gabon
TSA	TSZ	Tunisia
TTA	TTZ	Chad
TUA	TUZ	Ivory Coast
TVA	TXZ	France
TYA	TYZ	Benin
TZA	TZZ	Mali
UAA	UIZ	Russian Federation
UJA	UMZ	Uzbekistan
UNA	UQZ	Kazakhstan
URA	UZZ	Ukraine
V2A	V2Z	Antigua & Barbuda
V3A	V3Z	Belize
V4A	V4Z	St Kitts & Nevis
V5A	V5Z	Namibia
V6A	V6Z	Micronesia
V7A	V7Z	Marshall Is
V8A	V8Z	Brunei
VAA	VGZ	Canada
VHA	VNZ	Australia
VOA	VOZ	Canada
VPA	VSZ	United Kingdom
VTA	VWZ	India
VXA	VYZ	Canada
VZA	VZZ	Australia
WAA	WZZ	USA
XAA	XIZ	Mexico
XJA	XOZ	Canada
XPA	XPZ	Denmark
XQA	XRZ	Chile
XSA	XSZ	China
XTA	XTZ	Burkino Faso (Upper Volta)
XUA	XUZ	Kampuchea
XVA	XVZ	Vietnam
XWA	XWZ	Laos
XXA	XXZ	Portugal
XYA	XZZ	Burma (Myanmar)
Y2A	Y9Z	Germany
YAA	YAZ	Afghanistan
YBA	YHZ	Indonesia
YIA	YIZ	Iraq
YJA	YJZ	Vanuatu
YKA	YKZ	Syria
YLA	YLZ	Latvia
YMA	YMZ	Turkey
YNA	YNZ	Nicaragua
YOA	YRZ	Romania
YSA	YSZ	El Salvador
YTA	YUZ	Yugoslavia
YVA	YYZ	Venezuela
YZA	YZZ	Yugoslavia
Z2A	Z2Z	Zimbabwe
Z3A	Z3Z	Macedonia
ZAA	ZAZ	Albania
ZBA	ZJZ	United Kingdom
ZKA	ZMZ	New Zealand
ZNA	ZOZ	United Kingdom
ZPA	ZPZ	Paraguay
ZQA	ZQZ	United Kingdom
ZRA	ZUZ	South Africa
ZVA	ZZZ	Brazil
2AA	2ZZ	United Kingdom
3AA	3AZ	Monaco
3BA	3BZ	Mauritius
3CA	3CZ	Equatorial Guinea
3DA	3DM	Swaziland
3DN	3DZ	Fiji
3EA	3FZ	Panama
3GA	3GZ	Chile
3HA	3UZ	China
3VA	3VZ	Tunisia
3WA	3WZ	Vietnam
3XA	3XZ	Republic of Guinea
3YA	3YZ	Norway
3ZA	3ZZ	Poland
4AA	4CZ	Mexico
4DA	4IZ	Philippines

continued overleaf

continued from previous page

Prefix allocation From	*to*	*Country*	*Prefix allocation From*	*to*	*Country*
4JA	4KZ	Azerbaijan	6XA	6XZ	Madagascar
4LA	4LZ	Georgia	6YA	6YZ	Jamaica
4MA	4MZ	Venezuela	6ZA	6ZZ	Liberia
4NA	4OZ	Yugoslavia	7AA	7IZ	Indonesia
4PA	4SZ	Sri Lanka	7JA	7NZ	Japan
4TA	4TZ	Peru	7OA	7OZ	Republic of Yemen
4UA	4UZ	United Nations	7PA	7PZ	Lesotho
4VA	4VZ	Haiti	7QA	7QZ	Malawi
4WA	4WZ	Republic of Yemen	7RA	7RZ	Algeria
4XA	4XZ	Israel	7SA	7SZ	Sweden
4YA	4YZ	International Civilian Aviation Organisation	7TA	7YZ	Algeria
			7ZA	7ZZ	Saudi Arabia
			8AA	8IZ	Indonesia
4ZA	4ZZ	Israel	8JA	8NZ	Japan
5AA	5AZ	Libya	8OA	8OZ	Botswana
5BA	5BZ	Cyprus	8PA	8PZ	Barbados
5CA	5GZ	Morocco	8QA	8QZ	Maldives
5HA	5IZ	Tanzania	8RA	8RZ	Guyana
5JA	5KZ	Colombia	8SA	8SZ	Sweden
5LA	5MZ	Liberia	8TA	8YZ	India
5NA	5OZ	Nigeria	8ZA	8ZZ	Saudi Arabia
5PA	5QZ	Denmark	9AA	9AZ	Croatia
5RA	5SZ	Madagascar	9BA	9DZ	Iran
5TA	5TZ	Mauritania	9EA	9FZ	Ethiopia
5UA	5UZ	Niger Republic	9GA	9GZ	Ghana
5VA	5VZ	Togo	9HA	9HZ	Malta
5WA	5WZ	Western Samoa	9IA	9JZ	Zambia
5XA	5XZ	Uganda	9KA	9KZ	Kuwait
5YA	5ZZ	Kenya	9LA	9LZ	Sierra Leone
6AA	6BZ	Egypt	9MA	9MZ	Malaysia
6CA	6CZ	Syria	9NA	9NZ	Nepal
6DA	6JZ	Mexico	9OA	9TZ	Zaire
6KA	6NZ	Republic of Korea (South Korea)	9UA	9UZ	Burundi
			9VA	9VZ	Singapore
6OA	6OZ	Somali Republic	9WA	9WZ	Malaysia
6PA	6SZ	Pakistan	9XA	9XZ	Rwanda
6TA	6UZ	Sudan	9YA	9ZZ	Trinidad & Tobago
6VA	6WZ	Senegal			

Appendix B
Amateur radio prefixes

Prefix	*Country*	*CQ Zone*	*ITU Zone*
A2	Botswana	38	57
A3	Tonga	32	62
A4	Oman	21	39
A5	Bhutan	22	41
A6	United Arab Emirates	21	39
A7	Qatar	21	39
A9	Bahrain	21	39
AA–AG, AI–AK	United States		
AH1–AH0	*see* KH1–KH0		
AL7	Alaska KL7	01	01
AM	Spain EA	14	37
AP	Pakistan	21	41
BV	Taiwan	24	44
BV9P	Pratas Is		
BV9S	Spratley Archipelago		
BY	China	24	33
C2	Nauru	31	65
C3	Andorra	14	27
C5	Gambia	35	46
C6	Bahamas	08	11
C9	Mozambique	37	53
CE	Chile	12	14
CE0A,E,F	Easter Island	12	63
CE0X	San Felix	12	14
CE0Z	Juan Fernandez	12	14
CM,CO	Cuba	08	11
CN	Morocco	33	37
CP	Bolivia	10	12
CT3	Madeira Is	33	36

continued overleaf

continued from previous page

Prefix	*Country*	*CQ Zone*	*ITU Zone*
CT,CR,CS	Portugal	14	37
CU	Azores	14	36
CX	Uruguay	13	14
CY0	Sable Is	05	09
CY9	St Paul Is	05	09
D2	Angola	36	52
D4	Cape Verde	35	46
D6	Comoros	39	53
DA–DL	Germany	14	28
DU	Philippines	27	50
E3	Eritrea	37	48
EA	Spain	14	37
EA6	Balearic Is	14	37
EA8	Canary Is	33	36
EA9	Melilla	33	37
EA9	Ceuta	33	37
EI	Ireland	14	27
EK	Armenia	21	29
EL	Liberia	35	46
EM,EO	Ukraine	16	29
EP	Iran	21	40
ER	Moldova	16	29
ES	Estonia	15	29
ET	Ethiopia	37	48
EU–EW	Belarus	16	29
EX	Kyrghyzstan	17	30, 31
EY	Tadjikistan	17	30
EZ	Turkmenistan	17	30
F	France	14	27
FG	Guadeloupe	08	11
FH	Mayotte	39	53
FK	New Caledonia	32	56
FM	Martinique	08	11
FO	Tahiti	32	63
FO8X	Clipperton	07	10
FP	St Pierre & Miquelon	05	09
FR	Reunion	39	53
FR/G	Glorioso	39	53
FR/J	Juan De Nova	39	53
FR/T	Tromelin	39	53
FS	St Martin	08	11
FT8W	Crozet	39	68
FT8X	Kerguelen Is	39	68
FT8Z	Amsterdam Is & St Paul Is	39	68
FW	Wallis Is & Futuna Is	32	62
FY	French Guiana	09	12
G	England	14	27

continued overleaf

continued from previous page

Prefix	*Country*	*CQ Zone*	*ITU Zone*
GB	England (Special Event)	14	27
GC	Wales (Club Station)	14	27
GD	Isle of Man	14	27
GH	Jersey (Club Station)	14	27
GI	Northern Ireland	14	27
GJ	Jersey	14	27
GM	Scotland	14	27
GN	Northern Ireland (Club Station)	14	27
GP	Guernsey (Club Station)	14	27
GS	Scotland (Club Station)	14	27
GT	Isle of Man (Club Station)	14	27
GU	Guernsey	14	27
GW	Wales	14	27
GX	England (Club Station)	14	27
H2	Cyprus	20	39
H4	Solomon Is	28	51
HA,HG	Hungary	15	28
HB	Switzerland	14	28
HB0	Liechtenstein	14	28
HC	Ecuador	10	12
HC8	Galapagos	10	12
HH	Haiti	08	11
HI	Dominican Rep	08	11
HK	Colombia	09	12
HK0	San Andres Is	07	11
HK0	Malpelo Is	09	12
HL	South Korea	25	44
HP	Panama	07	11
HR	Honduras	07	11
HS	Thailand	26	49
HV	Vatican City	15	28
HZ	Saudi Arabia	21	39
I	Italy	15	28
IS0	Sardinia	15	28
IT9	Sicily	15	28
J2	Djibouti	37	48
J3	Grenada	08	11
J5	St Lucia	08	11
J7	Dominica	08	11
J8	St Vincent	08	11
JA,JE–JS	Japan	25	45
JD	Minami Torishima	27	90
JD	Ogasawara	27	45
JT	Mongolia	23	32
JW	Svalbard Is	40	18
JX	Jan-Mayen	40	18
JY	Jordan	20	39

continued overleaf

continued from previous page

Prefix	*Country*	*CQ Zone*	*ITU Zone*
K, KA–KZ	United States		
KC4	Antarctica	12	67
KC6	East Carolines	27	65
KC6	West Carolines (Belau)	27	64
KG4	Guantanamo Bay	08	11
KG6	Guam	27	64
KH0	Mariana Is	27	64
KH1	Baker Howland	31	61
KH2	Guam	27	64
KH3	Johnston Is	31	61
KH4	Midway Is	31	61
KH5	Palmyra Is	31	61
KH5K	Kingman Reef	31	61
KH6	Hawaii	31	61
KH7	Kure Is	31	61
KH8	American Samoa	32	62
KH9	Wake Is	31	65
KL7	Alaska	01	01
KP1	Navassa Is	08	11
KP2	Virgin Is	08	11
KP4	Puerto Rico	08	11
KP5	Desecheo Is	08	11
LA	Norway	14	18
LU	Argentina	13	14, 16
LX	Luxembourg	14	27
LY	Lithuania	15	29
LZ	Bulgaria	20	28
M	England	14	27
MC	Wales (Club Station)	14	27
MD	Isle of Man	14	27
MH	Jersey (Club Station)	14	27
MI	Northern Ireland	14	27
MJ	Jersey	14	27
MM	Scotland	14	27
MN	Northern Ireland (Club Station)	14	27
MP	Guernsey (Club Station)	14	27
MS	Scotland (Club Station)	14	27
MT	Isle of Man (Club Station)	14	27
MU	Guernsey	14	27
MW	Wales	14	27
MX	England (Club Station)	14	27
N	United States	04	07
NH1–NH0	*see* KH1–KH0		
NL7	Alaska	01	01
NP1–NP5	*see* KP1–KP5		
OA	Peru	10	12
OD	Lebanon	20	39

continued overleaf

continued from previous page

Prefix	*Country*	*CQ Zone*	*ITU Zone*
OE	Austria	15	28
OH	Finland	15	18
OH0	Aland Is	15	18
OJ0,OH0M	Market Reef	15	18
OK,OL	Czech Republic	15	28
OM	Slovak Republic	15	28
ON	Belgium	14	27
OX	Greenland	40	05, 75
OY	Faroe Is	14	18
OZ	Denmark	14	18
P2	Papua New Guinea	28	51
P4	Aruba	09	11
P5	North Korea	25	44
PA	Netherlands	14	27
PJ1,2,3,4,9	Netherlands Antilles	09	11
PJ5,6,7,8	St-Maarten	08	11
PY, PP–PW	Brazil	11	13
PY0	St Peter & Paul Rocks	11	13
PY0	Trindade Is	11	15
PY0	Fernando De Noronha	11	13
PZ	Surinam	09	12
R, RA–RZ	Russian Federation	15,16,17 18,19,23	19,20,21, 22,23,24,25,26,29, 30,31,32,33,34,35
S0	Western Sahara	33	46
S2	Bangladesh	22	41
S5	Slovenia	15	28
S7	Seychelles	39	53
S9	Sao Tome	36	47
SM	Sweden	14	18
SP	Poland	15	28
ST	Sudan	34	48
ST0	Southern Sudan	34	48
SU	Egypt	34	38
SV	Greece	20	28
SV/A	Mount Athos	20	28
SV5	Dodecanese	20	28
SV9	Crete	20	28
T2	Tuvalu	31	65
T30	West Kiribati	31	65
T31	Central Kiribati	31	62
T32	East Kiribati	31	61
T33	Banaba Is	31	65
T5	Somalia	37	48
T7	San Marino	15	28
T9	Bosnia Herzegovina	15	28
TA	Turkey	20	39

continued overleaf

continued from previous page

Prefix	*Country*	*CQ Zone*	*ITU Zone*
TF	Iceland	40	17
TG	Guatemala	07	11
TI	Costa Rica	07	11
TI9	Cocos Is	07	11
TJ	Cameroon	36	47
TK	Corsica	15	28
TL	Central Africa Rep	36	47
TN	Congo	36	52
TR	Gabon	36	52
TT	Chad	36	47
TU	Ivory Coast	35	46
TY	Benin	35	46
TZ	Mali	35	46
UA–UI	Russia		
UJ	Kazakhstan	17	30
UK	Uzbekistan	17	30
UR–UZ	Ukraine	16	29
V2	Antigua	08	11
V3	Belize	07	11
V4	St Kitts & Nevis	08	11
V5	Namibia	38	57
V6	Federation of Micronesia	27	65
V73	Marshall Is	31	65
V85	Brunei	28	54
VE,VO,VY	Canada	1,2,3,4,5	2,3,4,9 75
VK	Australia	29,30	55,58,59
VK0	Heard Is	39	68
VK0	Macquarie Is	30	60
VK9/M	Mellish Reef	30	56
VK9/N	Norfolk Is	32	60
VK9/W	Willis Is	30	55
VK9/L	Lord Howe Is	30	60
VK9/C	Cocos Keeling	29	54
VK9/X	Christmas Is	29	54
VP2E	Anguilla	08	11
VP2M	Montserrat	08	11
VP2V	British Virgin Is	08	11
VP5	Turks & Caicos	08	11
VP8	South Sandwich Is	13	73
VP8	South Georgia	13	73
VP8	Falkland Is	13	16
VP8	South Shetland	13	73
VP8	South Orkney	13	73
VP9	Bermuda	05	11
VQ9	Chagos	39	41
VR6	Pitcairn Is	32	63

continued overleaf

continued from previous page

Prefix	*Country*	*CQ Zone*	*ITU Zone*
VS6,VR2	Hong Kong	24	44
VU	India	22	41
VU4	Andaman & Nicobar Is	26	49
VU7	Laccadive Is (Lakshadweep)	22	41
W,WA–WZ	United States		
WH1–WH0	*see* KH1–KH0		
WL7	Alaska	01	01
XE	Mexico	06	10
XF4	Revilla Gigedo	06	10
XT	Burkina Faso	35	46
XU	Kampuchea	26	49
XW	Laos	26	49
XX9	Macao	24	44
XZ	Burma (Myanmar)	26	49
YA	Afghanistan	21	40
YB–YH	Indonesia	28	51
YI	Iraq	21	39
YJ	Vanuatu	32	56
YK	Syria	20	39
YL	Latvia	15	29
YN	Nicaragua	07	11
YO	Romania	20	28
YS	El Salvador	07	11
YU,YT	Serbia	15	28
YV	Venezuela	09	12
YV0	Aves Is	08	11
Z2	Zimbabwe	38	53
Z3	Macedonia	15	28
ZA	Albania	15	28
ZB	Gibraltar	14	37
ZC4	Cyprus - UK Sovereign Base	20	39
ZD7	St Helena	36	66
ZD8	Ascension Is	36	66
ZD9	Tristan Da Cunha & Gough Is	38	66
ZF	Cayman Is	08	11
ZK1	South Cook Is	32	62
ZK1	North Cook Is	32	62
ZK2	Niue Is	32	62
ZK3	Tokelau Is	31	62
ZL	New Zealand	32	60
ZL5	New Zealand bases in Antarctica	12	67
ZL7	Chatham Is	32	60
ZL8	Kermadec Is	32	60
ZL9	Auckland & Campbell Is	32	60
ZP	Paraguay	11	14
ZS	South Africa	38	57
ZS8	Marion Is & Prince Edward Is	38	57

continued overleaf

continued from previous page

Prefix	*Country*	*CQ Zone*	*ITU Zone*
ZV–ZZ	Brazil	11	13
1A	Sovereign Military Order of Malta (Rome)	15	28
1P	Seborga	15	28
1S	Spratly Is	26	50
2D	Isle of Man (Novices)	14	27
2E	England (Novices)	14	27
2I	Northern Ireland (Novices)	14	27
2J	Jersey (Novices)	14	27
2M	Guernsey (Novices)	14	27
2W	Wales (Novices)	14	27
3A	Monaco	14	27
3B6	Agalega	39	53
3B8	Mauritius	39	53
3B9	Rodriguez Is	39	53
3C	Equatorial Guinea	36	47
3C0	Annobon (Pagalu)	36	52
3D2	Conway Reef	32	56
3D2	Rotuma	32	56
3D2	Fiji Is	32	56
3DA0	Swaziland	38	57
3V	Tunisia	33	37
3W	Vietnam	26	49
3X	Guinea	35	46
3Y	Bouvet Is	38	67
3Y	Peter Is	12	72
4J	Azerbaijan	21	29
4L	Georgia	21	29
4S	Sri Lanka	22	41
4U	ITU Geneva	14	28
4W	Yemen	21	39
4X,4Z	Israel	20	39
5A	Libya	34	38
5B	Cyprus	20	39
5H	Tanzania	37	53
5N	Nigeria	35	46
5R	Malagasy Rep (Madagascar)	39	53
5T	Mauritania	35	46
5U	Niger	35	46
5V	Togo	35	46
5W	Western Samoa	32	62
5X	Uganda	37	48
5Z	Kenya	37	48
6W	Senegal	35	46
6Y	Jamaica	08	11
7P	Lesotho	38	57
7Q	Malawi	37	53

continued overleaf

continued from previous page

Prefix	*Country*	*CQ Zone*	*ITU Zone*
7X	Algeria	33	37
7Z	Saudi Arabia	21	39
8P	Barbados	08	11
8Q	Maldive Is	22	41
8R	Guyana	09	12
9A	Croatia	15	28
9G	Ghana	35	46
9H	Malta	15	28
9J	Zambia	36	53
9K	Kuwait	21	39
9L	Sierra Leone	35	46
9M0	Spratly Is	26	50
9M2	West Malaysia	28	54
9M4	West Malaysia	28	54
9M6	East Malaysia	28	54
9M8	East Malaysia	28	54
9N	Nepal	22	42
9Q	Zaire	36	52
9U	Burundi	36	52
9V	Singapore	28	54
9X	Rwanda	36	52
9Y	Trinidad	09	11

Appendix C
Call areas

United States

Number in Prefix	*State*	*CQ Zone*	*ITU Zone*
1	Connecticut	05	08
	Maine	05	08
	Massachusetts	05	08
	New Hampshire	05	08
	Rhode Island	05	08
	Vermont	05	08
2	New Jersey	05	08
	New York	05	08
3	Delaware	05	08
	Washington DC	05	08
	Maryland	05	08
	Pennsylvania	05	08
4	Alabama	04	08
	Florida	05	08
	Georgia	05	08
	Kentucky	04	08
	North Carolina	05	08
	South Carolina	05	08
	Tennessee	04	08
	Virginia	05	08
5	Arkansas	04	08
	Louisiana	04	08
	Mississippi	04	08
	New Mexico	04	07
	Oklahoma	04	07
	Texas	04	07
6	California	03	06

continued overleaf

continued from previous page

Number in Prefix	*State*	*CQ Zone*	*ITU Zone*
7	Arizona	03	06
	Idaho	03	06
	Montana	04	07
	Nevada	03	06
	Oregon	03	06
	Utah	04	07
	Washington	03	06
	Wyoming	04	07
8	Michigan	04	08
	Ohio	04	08
	West Virginia	05	08
9	Illinois	04	08
	Indiana	04	08
	Wisconsin	04	08
0	Colorado	04	07
	Iowa	04	07
	Kansas	04	07
	Minnesota	04	07
	Missouri	04	07
	Nebraska	04	07
	North Dakota	04	07
	South Dakota	04	07

Canada

Prefix	*State*	*CQ Zone*	*ITU Zone*
VE1	New Brunswick	05	09
VO1	Newfoundland	05	09
VE1	Nova Scotia	05	09
VE2	Quebec	05	04
VO2	Labrador	02	04
VY2	Prince Edward Is	05	04
VE3	Ontario	04	04
VE4	Manitoba	04	03
VE5	Saskatchewan	04	03
VE6	Alberta	03	02
VE7	British Columbia	03	02
VE8	Yukon	01	75
VY1	Yukon	01	75

Index